The Politics of Nuclear Energy in China

Energy, Climate and the Environment Series

Series Editor: **David Elliott**, Emeritus Professor of Technology, Open University, UK

Titles include:

David Elliott *(editor)*
NUCLEAR OR NOT?
Does Nuclear Power Have a Place in a Sustainable Future?

David Elliott *(editor)*
SUSTAINABLE ENERGY
Opportunities and Limitations

Horace Herring and Steve Sorrell *(editors)*
ENERGY EFFICIENCY AND SUSTAINABLE CONSUMPTION
The Rebound Effect

Matti Kojo and Tapio Litmanen *(editors)*
THE RENEWAL OF NUCLEAR POWER IN FINLAND

Antonio Marquina *(editor)*
GLOBAL WARMING AND CLIMATE CHANGE
Prospects and Policies in Asia and Europe

Catherine Mitchell
THE POLITICAL ECONOMY OF SUSTAINABLE ENERGY

Ivan Scrase and Gordon MacKerron *(editors)*
ENERGY FOR THE FUTURE
A New Agenda

Gill Seyfang
SUSTAINABLE CONSUMPTION, COMMUNITY ACTION AND THE NEW
ECONOMICS
Seeds of Change

Joseph Szarka
WIND POWER IN EUROPE
Politics, Business and Society

Xu Yi-chong
THE POLITICS OF NUCLEAR ENERGY IN CHINA

Energy, Climate and the Environment
Series Standing Order ISBN 978–0–230–00800–7 (hb) 978–0–230–22150–5 (pb)

You can receive future titles in this series as they are published by placing a standing order.
Please contact your bookseller or, in case of difficulty, write to us at the address below with your
name and address, the title of the series and the ISBN quoted above.

Customer Services Department, Macmillan Distribution Ltd, Houndmills, Basingstoke,
Hampshire RG21 6XS, England

The Politics of Nuclear Energy in China

Xu Yi-chong
Research Professor, Griffith University, Australia

© Xu Yi-chong 2010

First published 2010 by
PALGRAVE MACMILLAN

Palgrave Macmillan in the UK is an imprint of Macmillan Publishers Limited, registered in England, company number 785998, of Houndmills, Basingstoke, Hampshire RG21 6XS.

Palgrave Macmillan in the US is a division of St Martin's Press LLC, 175 Fifth Avenue, New York, NY 10010.

Palgrave Macmillan is the global academic imprint of the above companies and has companies and representatives throughout the world.

Palgrave® and Macmillan® are registered trademarks in the United States, the United Kingdom, Europe and other countries.

ISBN: 978–0–230–22890–0 hardback

This book is printed on paper suitable for recycling and made from fully managed and sustained forest sources. Logging, pulping and manufacturing processes are expected to conform to the environmental regulations of the country of origin.

A catalogue record for this book is available from the British Library.

Library of Congress Cataloging-in-Publication Data

Xu, Yi-Chong.
    The politics of nuclear energy in China / Xu Yi-chong.
        p. cm.—(Energy, climate, and the environment)
    Includes bibliographical references.
    ISBN 978–0–230–22890–0
        1. Nuclear energy—Government policy—China. 2. Nuclear industry—China. I. Title.

HD9698.C62X8 2010
333.792'40951—dc22   2010027572

10 9 8 7 6 5 4 3 2 1
19 18 17 16 15 14 13 12 11 10

Printed and bound in Great Britain by
CPI Antony Rowe, Chippenham and Eastbourne

Contents

Illustrations

Tables

Figures

Abbreviations

ADB	Asian Development Bank
AECL	Atomic Energy of Canada Limited
ASE	Atomstroy-export
BNFL	British Nuclear Fuels plc
BWR	Boiling water reactor
CAEA	China Atomic Energy Authority
CGNPC	China Guangdong Nuclear Power Corporation
CIAE	China Institute of Atomic Energy
CNEC	China Nuclear Engineering and Construction Corporation
CNEIC	China Nuclear Energy Industry Corporation
CNNC	China National Nuclear Corporation
CNPC	China National Petroleum Corporation
COSTIND	Commission of Science, Technology and Industry for National Defence
CPI	China Power Investment Corporation
CSRC	China Securities Regulatory Commission
CTIEC	China Technology Import and Export Corporation
ECGD	Export Credit Guarantee Department
EDF	Electricite de France
EPR	European pressurised reactor
FDI	Foreign direct investment
GDP	Gross domestic product
GEC	General Electric Company
GHG	Greenhouse gas
GNPJV	Guangdong Nuclear Power Joint Venture Company
GW	Gigawatt
GWe	Gigawatt electric
HEU	Highly enriched uranium
HKNIC	Hong Kong Nuclear Investment Corporation
HTR	High-temperature gas-cooled reactor
HTR-PM	High-temperature gas-cooled reactor pebble-bed module
IAEA	International Atomic Energy Agency
IEA	International Energy Agency
INET	Institute of Nuclear and New Energy Technology
IPSN	Instiut de Protection et de Sureté Nucléaire

kgU	Kilograms of uranium
Kwh	Kilowatt hour
LEU	Low enriched uranium
LWR	Light water reactor
MEP	Ministry of Environmental Protection
MIIT	Ministry of Industry and Information Technology
Minatom	Russian Ministry of Atomic Energy
MLR	Ministry of Land and Resources
MNI	Ministry of Nuclear Industry
MOU	Memorandum of Understanding
MOX	Mixed oxide fuel
MW	Megawatt
MWREP	Ministry of Water Resources and Electrical Power
NDRC	National Development and Reform Commission
NEA	Nuclear Energy Agency (OECD)
NEA	National Energy Administration
NGO	Non-governmental organisation
NNSA	National Nuclear Safety Administration
NPC	National People's Congress
NPP	Nuclear power plant
NPT	Nuclear Non-Proliferation Treaty
NRC	Nuclear Regulatory Commission
O&M	Operation and management
OECD	Organisation for Economic Cooperation and Development
PHWR	Pressurised heavy water reactor
PWR	Pressurised water reactor
R&D	Research and development
RAR	Reasonably assured resources
RMB	Renminbi
SASAC	State-owned Asset Supervision and Administration Commission
SERC	State Electricity Regulatory Commission
SETC	State Economic and Trade Commission
Sino-Uranium	China Nuclear International Uranium Corporation
SNEDRI	Shanghai Nuclear Engineering and Design Research Institute
SNPTC	State Nuclear Power Technology Corporation Ltd
SPC	State Planning Commission
SPCC	State Power Corporation of China

SSTC	State Science and Technology Commission
SWU	Separative work units
tU	Tonnes of uranium
US	United States
USEC	United States Enrichment Corporation

Preface and Acknowledgements

This book is about nuclear energy and its development in China. It is not a book about nuclear weapons programmes. There are many books available on nuclear weapons programmes and on the issue of non-proliferation. Nuclear power as a source of energy in developing countries, however, has seldom been studied. Nuclear energy interested me initially because it is believed in China and some quarters of the international community that nuclear energy is the only clean and reliable source of electricity that can be used to meet large baseload demand. If the nuclear energy programme is successful, so the argument runs, it may make some difference to the energy mix and alleviation of environmental pollution in China. This project sought to see if it could.

The project has turned out to be interesting, not only in terms of the complex issues involved in nuclear energy development, but also because the politics proved to be counterintuitive – China, nominally a centralised authoritarian regime, has no single central authority to drive nuclear energy or develop strategies. Nuclear policy, notionally the most secretive of policy areas, has been openly debated and publicly contested on several fronts: whether a project should be launched; who should pay for what; what and whose technology should be adopted; and whose interests should be taken into consideration – those who would benefit from the jobs created by a project and the electricity to be generated, or those who did not want to see the ugly scenery of nuclear power plants or were afraid of potential accidents. The politics, so surprising and so diverse, made this project a most enjoyable exercise.

Many people have helped with this project. Greg Burke is a wonderful research assistant, patient, meticulous and effective. James Arklay, a young man with great potential, assisted in some of the translation and read the manuscript with great care. Keith Whittam and Paula Cowan edited the text with considerable sensitivity. Alexandra Webster at Palgrave assisted the whole process.

The greatest appreciation goes to Pat Weller who listened patiently, debated and discussed the ideas throughout this whole project, and read the drafts over and over again when he was trying to meet his own book deadlines. More importantly, he is always ready to help at the most critical moments in my writing or thinking, and always has time for a word of encouragement.

This project has been made possible with the financial support from the Australian Research Council.

Xu
Brisbane
1 February 2010

Series Editor's Preface

Energy, Climate and the Environment

Concerns about the potential environmental, social and economic impacts of climate change have led to a major international debate over what could and should be done to reduce emissions of greenhouse gases, which are claimed to be the main cause. There is still a scientific debate over the likely scale of climate change, and the complex interactions between human activities and climate systems, but, in the words of no less than the Governor of California, Arnold Schwarzenegger, *"I say the debate is over. We know the science, we see the threat, and the time for action is now."*

Whatever we now do, there will have to be a lot of social and economic adaptation to climate change, preparing for increased flooding and other climate related problems. However, the more fundamental response is to try to reduce or avoid the human activities that are seen as causing climate change. That means, primarily, trying to reduce or eliminate emission of greenhouse gasses from the combustion of fossil fuels in vehicles and power stations. Given that around 80% of the energy used in the world at present comes from these sources, this will be a major technological, economic and political undertaking. It will involve reducing demand for energy (via lifestyle choice changes), producing and using whatever energy we still need more efficiently (getting more from less), and supplying the reduced amount of energy from non-fossil sources (basically switching over to renewables and/or nuclear power).

Each of these options opens up a range of social, economic and environmental issues. Industrial society and modern consumer cultures have been based on the ever-expanding use of fossil fuels, so the changes required will inevitably be challenging. Perhaps equally inevitable are disagreements and conflicts over the merits and demerits of the various options and in relation to strategies and policies for pursuing them. These conflicts and associated debates sometimes concern technical issues, but there are usually also underlying political and ideological commitments and agendas that shape, or at least colour, the ostensibly technical debates. In particular, at times, technical assertions can be used to buttress specific policy frameworks in ways that subsequently prove to be flawed

The aim of this series is to provide texts that lay out the technical, environmental and political issues relating to the various proposed policies for responding to climate change. The focus is not primarily on the science of climate change, or on the technological detail, although there will be accounts of the state of the art, to aid assessment of the viability of the various options. However, the main focus is the policy conflicts over which strategy to pursue. The series adopts a critical approach and attempts to identify flaws in emerging policies, propositions and assertions. In particular, it seeks to illuminate counter-intuitive assessments, conclusions and new perspectives. The aim is not simply to map the debates, but to explore their structure, their underlying assumptions and their limitations. Texts are incisive and authoritative sources of critical analysis and commentary, indicating clearly the divergent views that have emerged and also identifying the shortcomings of these views. However the books do not simply provide an overview, they also offer policy prescriptions.

There are many conflicting views of how best to proceed in order to respond to climate change and the security threats it may involve, for example as to whether adaptation is more urgent than mitigation, and over who should take the initiative and pay for them. There are also uncertainties over how countries might best prepare to cope with climate threats and deal with the social, economic and political conflicts that could emerge, both nationally and internationally, as climate impacts increase. Opinions differ in part because the intensity of the impacts on each area may differ. And the level of commitment varies, as of course does the capacity to deal with the threats. However, what seems to be emerging is a view that, since climate change will effect everyone to some degree, directly or indirectly, common global solutions are needed. What remains in doubt is whether agreement can be reached on these solutions in time to avoid major social and economic problems, and be achieved without requiring or leading to the adoption of divisive, inequitable or authoritarian policies.

1
Introduction

China is planning to increase its nuclear generation capacity by building two or three nuclear power plants every year until at least 2020, as one step to meet its rapidly rising energy demands and mitigate climate change threats. Will China be able to expand its nuclear power generation capacity in sufficient quantities and with sufficient speed to beat the urgent twin challenges it faces – energy security and climate change? This book seeks to understand *the constellation of political forces in China that has shaped its nuclear energy development.*

Economy, energy and environment are inextricably linked. A secure energy supply is needed to fuel economic growth; energy production and consumption, particularly combustion of fossil fuels, produce environmental pollution, and this pollution has direct impacts and social costs. These challenges have become more pressing in the past three decades as developing countries enter the stage of industrialisation and urbanisation and struggle to provide their citizens with modern energy – electricity and heat. Providing adequate and reliable energy supplies to meet the demand for economic and social development while controlling or even reducing greenhouse gas (GHG) emissions is necessary for sustainable development. Eliminating or mitigating the effects of pollution, or moving to non-polluting energy sources also entails economic costs and requires changes in personal, community, corporate and national behaviour.

These challenges are not just technical or economic; nor are they unique to any single country. Energy is a political issue. On one side of the political spectrum, the answer to the energy problems rests on a continuation of globalisation based on free market principles – free markets will ensure adequate investment capital to meet rising energy demands[1] and free markets can also change consumers' behaviour to deal with environmental problems. On the other side of the spectrum is the argument for government intervention and regulation: energy is essential to the national economy and living standards of the people and therefore governments should decide how capital and energy supplies are apportioned and allocated.

More importantly, energy issues often pit one group of interests against another.[2] Energy companies stand to benefit from high energy prices while end-users affected by the high price of inputs may lose their margins of profit or even their livelihood. As energy end-users, we like to maintain and improve the comfort of our living (driving SUVs and keeping electric appliances on stand-by) but do not want to see power plants, hydro dams, transmission grids or uranium disposal sites in our backyards. While development of renewable energy is urgent and needs more resources, traditional fossil fuels sectors are unwilling to give up their dominant position. These are only a few examples of the difficult political issues. Governments have to find the right balance between energy and economy, energy and development, energy and environment, energy and security, and, more importantly, clean energy and adequate and reliable energy supplies. To understand similar challenges that China is facing, however, we need to consider all these difficulties, multiply by the number of its population, and square it with low GDP per capita and poor natural resources endowment.[3]

After three decades of a near double-digit annual growth rate, China had achieved what seemed impossible: over 500 million people had been lifted from absolute poverty; more than 268 million people had entered into the urban domain; its economy had become the second largest in the world in purchasing power parity terms; it had become the third world's largest trader; and in less than a generation, China moved "from being a minor and largely self-sufficient energy consumer to become the world's fastest-growing energy consumer and major player on the global energy market.'[4] China is also facing the most difficult and 'unique energy security challenge'[5] – to sustain rapid economic development and growth in output in a way that is more equitable, more environmentally sustainable and, therefore, less energy-intensive than has so far been the case.

These 'difficult-to-reconcile' objectives become more challenging when China faces the 'extremely fundamental forces' in its development – that is, China is like "a group of relatively developed islands with a cumulative population of over 400 million people that are scattered around in a sea of over 800 million people who live very much in developing-country conditions.'[6] To accommodate about 15 million people moving into urban areas each year, 'effectively, China has to build urban infrastructure and create urban jobs for a new, relatively poor city of 1.25 million people every month, and that will likely continue for the better part of the next two decades.'[7] Given that the per capita energy consumption for urban citizens is 3.5 times that of rural citizens, demands for energy and other infrastructure in China have placed tremendous pressure on resources and the environment, especially in the past decade.

By 2002, China had to stare at the dark side of its double-digit growth. Blackouts rolled in and factory lights flickered; the grid sucked dry by a decade of breakneck industrialisation. Oil and natural gas were running low. China's increasing import of oil triggered global panic in oil supplies in the mid-decade. On the home front, belching power plants were burning coal faster than cracked old railroads could deliver it. The most populous country on the planet was passing the US to become the largest GHG emitter and the home of 20 of the world's 30 worst polluted cities. A 2007 World Bank report, 'The Cost of Pollution in China: Economic Estimates of Physical Damage', estimates the total cost of air and water pollution as about 5.8% of its GDP annually; other scholars raised the direct cost of pollution damage to China's economy to between 8% and 13%. In human costs, an estimated 750,000 people die in China of pollution-related illness every year. Pollution is causing problems for the government as the frequency and size of protests over local environmental conditions increase.[8]

Chinese leaders encounter fundamental problems regarding human capital, infrastructure, social malaise, technical capabilities and, more importantly, institutional capabilities when facing several extremely fundamental facts of life in its development: (a) urbanisation and industrialisation will continue; (b) energy consumption will keep rising; (c) coal will remain the main source of its energy supply and (d) CO_2 emissions will keep rising too. Yet, some balance needs to be found and serious measures have to be taken to avoid one of 'the high-probability failures' – 'a power supply failure': a situation that is akin to the car running out of gas either because of the inability to secure energy supplies or an environmental collapse.[9]

To manage these twin challenges of 'energy security' and 'climate change', China seeks a reliable fuel supply from an array of sources, grasping at every energy alternative within its reach, including the traditional fossil fuels (coal, oil and natural gas), renewable resources (wind, solar, biomass, hydro and other forms), and nuclear. Development of renewable and nuclear energy will, and can, only supplement part of the increase in energy demands and its success depends on how the politics plays out.

The state of play

The human being has always sought 'to supplement [its] puny muscles by whatever means [it] could find: horses, elephants, tides, wind, water-mills, slaves.'[10] The discovery of coal and its usage in the 18th century brought about steam engines and made industrialisation possible; it also allowed a society to feed itself on the produce of foreign farms. The discovery of oil in the late 19th century and its wide usage in the 20th century were the preconditions for the industrialisation of North America and then around the world. Finding new energy sources has always been part of the human endeavour and controlling energy resources and their production is a necessary condition for a modern economy. Energy, however, is not used for its own sake, but rather for the services it makes possible – lighting, cooking, heating, cooling, generating electricity, and transporting freight and people.

There are different means of providing a desired service, each with its own costs and benefits. People in industrialised countries and the wealthy in developing countries use gas and electricity for lighting, cooking and heating while the poor burn charcoal, agricultural residues or animal dung to cook meals or to warm up their homes.

Providing people with access to modern energy – electricity and heat – has been considered a key to social and economic development in all countries by most politicians and all international organisations. As early as the 1930s, farmers in Texas were telling their Congressman, Lyndon Johnson, later the US President, about their life without electricity:

> Living was just drudgery then... Living – just *living* – was a problem. No lights. No plumbing. Nothing. Just living on the edge of starvation. That was farm life for us. God, city people think there was something fine about it. If they only knew. ...[11]

'I'll get it for you,' had promised Johnson. When people in Hill County, Texas, finally got electricity, 'people began to name their kids for Lyndon Johnson.'[12] This was what the Chinese government promised to do when reform started in the late 1970s – building up electricity infrastructure to support economic growth and thereby improve the standard of living of its people.

China's total electricity consumption quadrupled between 1980 and 2000 (from 259TWh to 1081TWh). It then grew by 14% per year between 2000 and 2007 and reached 2717TWh in 2007.[13] Over 400 million people in the country were connected to electricity and its consumption per capita rose from a little over 307kwh in 1980 to 2328kwh in 2007, which remained below the world's average (2752kwh) and still only 27% of that in OECD countries (8477kwh). Total installed generation capacity expanded from 66GW in 1980 to 316GW in 2000 and then more than doubled in the next 7 years to 706GW in 2007. It is expected to double again to 1460GW by 2020.[14]

Given that China is poor in other energy resources, coal meets about 70% of its total energy consumption. Coal-fired thermal generation plants produce 82% of its electricity and they are expanding too. In 2008, the total coal-fired thermal generation capacity under construction in China (112GW) was twice Australia's current total capacity from all sources (52GW). Heavy reliance on coal has had two immediate consequences: (a) coal is depleting quickly (see Chapter 3), and (b) 'burning coal contributes to 90% of the national total sulphur dioxide (SO_2) emissions, about 70% of the national total dust, nitrogen oxide (NO_x) emissions and carbon dioxide (CO_2) emissions'[15] (see Chapter 8). Both issues have raised serious concerns in China as well as in the international community.

The Chinese government has initiated a wide range of measures related to energy and climate change. It has specifically emphasised its intent to improve energy efficiency and lower energy intensity – the amount of energy used per unit of GDP. In 2002, for example, China announced that it aimed at quadrupling its economy while only doubling its energy use by 2020. In 2004, the State Council approved the Medium- and Long-term Energy Development Plan Outline 2004–2020 and the National Development and Reform Commission (NDRC) formulated the Medium- and Long-term Energy Conservation plan in the same year. In 2005, the Renewable Energy Law was adopted that particularly emphasised diversifying energy mix by expanding the share of renewable energy to 16% of China's total energy production by 2020.

China's 11th Five-Year Plan (FYP) (2006–10) was formulated in 2003–05 when it had become clear that capital-intensive and industry-led growth had placed tremendous pressures on energy, natural resources and the environment. The 11th FYP gave priority to rebalancing the economic structure and to environmental and social objectives. It set an ambitious target for energy-efficiency improvement: China would double its 2000 per capita GDP and reduce energy intensity of GDP by 20% by 2010. To achieve this ambitious objective, China would have to adopt structural, technical and managerial changes: '(i) *structural*, resulting from rebalancing the economic and industrial structure, particularly reducing the share of energy intensive industries; (ii) *technical*, through technical progress to reduce energy consumption per unit of product; and (iii) *managerial*, by reducing energy waste during energy production, transportation, and consumption through strengthening regulatory and administrative institutional capacity.'[16] Restructuring its economy, especially moving away from energy-intensive industries, such as steel, aluminium or petrochemical, to service sectors takes time and involves heavy social, economic and even political costs. It has never happened in any country without radical changes.[17]

Other specific policies include: improving energy efficiency, promoting clean and renewable energy, closing down small-sized and highly polluting power plants, building super-critical power plants (with capacity of 600MW plus), and developing a coal-bed methane industry as a way to use clean coal technology. In 2007, for example, it closed 533 small coal-fired power plants with a capacity of 21.6GW. It also closed small-sized steel and iron mills with a capacity of 84 billion tonnes, exceeding the planned 65 billion tonne target. By the end of 2007, 74% of ordered thermal capacity is 600MW and above.[18] The Chinese government also adopted policies to encourage the development of renewable energy.

In September 2007, a target was set for the five major power generation companies that at least 3% of their generation capacity would have to be from renewable resources by 2020. Wind and solar generation capacity has expanded much faster than anticipated. By 2008, over 60% of the world's solar water heaters were used in China. Most targets for renewable energy set in the early 2000s had already been exceeded by the end of 2008 (see Table 1.1).

Wind power generation capacity in China, especially, expanded at a much faster speed than anticipated. The first wind farm was completed and connected to the grid as early as 1986, but its development was very slow.[19] According to the Global Wind Energy Council, the total installed wind capacity increased from 346MW in 2000 to 12210MW in 2008 (see Table 1.2).

Given that the availability rate of wind power is only 20% and wind mills tend to locate in remote areas, there are technical barriers to connecting wind power to the grids. Consequently, a share of generation capacity does not translate into the same share of electricity production. For example, wind power accounted for 1.1% of the total installed generation capacity in 2008 in China and produced only 0.4% of its electricity.[20] According to NDRC, even if the target of 'non-hydro renewables (over 60GW by 2020) is achieved by 2020, [it] will represent a small fraction of total installed capacity for many years to come.'[21]

In addition to renewable energy (wind, solar and biomass), nuclear energy has also been advocated and promoted by the Chinese government as an alternative to meet rising demand and to deal with the environmental

Table 1.1 Renewable energy targets

	2010[a]	2020[a]	End of 2008[b]
Proportion of RE in national energy mix (%)	10	16	9
Hydro (GW)	180	300	170
Biomass (GW)	5.5	30	3.15
Wind (GW)	5	30	12.2
Solar (MW)	500	1800	150
Solar collection (million m^3)	150	300	125
Biogas (billion m^3)	15	30	14
Liquid fuel (million tonnes)	2	10	1.65

Source: [a]Baker & McKenzie, 'RELaw Assist: Renewable Energy Law in China – Issue Paper', p. 21, at http://www.bakernet.com/NR/rdonlyres/B06FB192-EF10–4304-B966-FBDF1A076A8C/0/relaw_issues_paper_jun07.pdf, June 2007.
[b]中国国家统计局，系列报告之十三：能源生产能力大幅提高结构不断优化， 22 September 2009 (China Statistical Bureau, 'Report Series No.13: 'Increased Energy Production with Improved Industrial Structure', 22 September 2009).

Table 1.2 Total installed wind capacity in China, 2000–08

Year	2000	2001	2002	2003	2004	2005	2006	2007	2008
MW	346	402	469	567	764	1260	2599	5910	12210

Source: Global Wind Energy Council, 'China', at http://www.gwec.net/index.php?id=125

pollution from burning coal. China's 10th FYP (2000–05) included the construction of 8 nuclear power plants and the 11th FYP (2006–10) proposed the construction of 14 new reactors. In 2005, the State Council adopted the Medium- to Long-term Nuclear Energy Development Plan, which specified the target of building 40GWe nuclear capacity by 2020 with another 18GW under construction. In 2009, the target was upgraded to 60–70GWe and more than 16 provinces, regions and municipalities announced intentions to build nuclear power plants to increase nuclear electricity production. By the end of 2009, China hosted 20 out of 54 nuclear power stations under construction in the world.

China has decided to expand its nuclear energy capacity. Many in the country believe nuclear is the future for China and the world because it is clean and has zero CO_2 emissions and near zero emission of other GHGs. It can meet base-load demand in areas where population is dense and demand is high. This position is supported by the International Energy Agency (IEA) that suggested nuclear energy expansion would alleviate the impending global energy crisis, reduce energy vulnerability, ease the impact of rising fossil-fuel prices, and mitigate GHG emissions.[22] It is also endorsed by the International Atomic Energy Agency (IAEA) that argues nuclear energy is an option, especially in places where: (a) energy demands grow rapidly; (b) alternative sources are scarce and expensive and (c) a nuclear energy programme is already in place.[23]

Understanding the nuclear development in China

Is nuclear energy indeed the future? The government in Beijing seems to think so and is putting a lot of resources into it. It eyes the nuclear industry in France, Japan and especially South Korea as examples. Nuclear energy produces 75% of the electricity in France, 29% in Japan, 20% in Taiwan and 35% in South Korea. In China, nuclear power plants produced only 1.9% of the country's electricity. China has had its nuclear energy programme since late 1970s and by the end of 2009, there were eleven reactors at six nuclear power stations in operation with an installed capacity of 9.1GWe – approximately 1.3% of the country's total installed generation capacity. Twenty-one units with capacity of over 20GWe were under construction and 24 units had been approved for construction, totalling 25.4GWe of capacity. Nuclear development is inseparable from the support of a nuclear fuel cycle industry and currently China has mastered the key technologies with uranium prospecting, mining, milling, conversion, enrichment and fuel fabrication. A dual objective in nuclear technology has been set up to adopt standardised technology for long-term nuclear development and to develop a home-based technology. It is nonetheless important to note that even if China manages to expand its nuclear generation capacity to 40GWe by 2020, it would still only account for 2.7% of its total generation capacity and produce less than 5% of the country's total electricity.[24]

Can China do it? Does the country have the political, economic, technical and human capacity to make nuclear power a viable option? History shows what China is capable of: it detonated nuclear bombs when it had a high degree of poverty, political instability and international isolation. It had gathered a small group of scientists, provided them with the necessary financial and other resources, and rallied its people behind the programme.[25] Now the country is facing different challenges: rapid transformation from an agriculture-based economy to an industrialised one and corresponding urbanisation are having a significant impact on the quantity and quality of energy resources available for production process and consumption, and on the ability of the environment to absorb the waste by-products deposited in the air, water and soil.

Adding to the urgency is the real and substantial depletion of global low-cost fossil fuel resources. China's aggressive search for adequate energy supplies around the world is causing widespread concerns. Consequently, nuclear power is seen as crucially important to help reduce its energy vulnerability and the risks of global climate change. Can it apply the same dedication and build more capacities to engage in an extensive nuclear energy programme that raises different challenges from nuclear weaponry?

The experience of nuclear energy development in other countries provides a cautionary tale: nuclear energy development often faces serious political opposition from both the left and the right,[26] high costs that no one is willing to absorb,[27] changing technology that even scientists cannot agree on,[28] and risks of waste management and proliferation that scare and deter many countries from undertaking the project.[29] Has the nuclear energy development in China met similar political challenges? What support and opposition has China confronted? What can its nuclear development, especially the transformation from a weapons programme to an energy one, tell us about the politics involved, and about the political, economic, technical and human capacity that will be needed for the expansion of nuclear energy as a clean energy source on a sustainable basis?

China's own record of nuclear energy leads to another cautionary tale. The nuclear policies have been inconsistent, contested and fragmented. In the past 30 years, various targets have been set up: in the early 1980s, the government wanted to build 10GWe nuclear generating capacity by 2000; in the early 1990s, the target of 18GWe was set for 2010; it was followed by 36GWe by 2020, which was upgraded to 40GWe in 2007. This repeated failure to meet targets is in contrast to most sectors of the economy, including renewable energy, where development has far exceeded targets set along the way in the past 30 years.

Is this failure of nuclear energy development the result of high economic risks and inadequate financial resources, which are often argued to be the determining factors for nuclear futures? Are 'economics and the comparative long-term costs of alternatives more likely to drive the politics'[30] in nuclear energy development, as some argue? Or are nuclear-related economic issues shaped more by politics than economic consideration, especially in transition economies? Would this economic calculation make any difference in developing and transition economies? Who should pay for energy investment and in what form? Have investment sources been different in nuclear energy from other energy sectors? If so, why?

Nuclear energy technology has been developed and debated ever since scientists extracted energy from the fission of uranium in 1939 – fission or fusion, using uranium or thorium, pressurised water reactor or boiling water reactor, or a full nuclear fuel cycle or only the once-through technology. The technical debate is beyond the discussion in this book; the choice of technology, however, is a political issue. What has been driving the decision on technological selection and adoption in China? Who is involved in the decision making and what are their interests? The issue on technology selection, more precisely the standardisation of reactors, is at the core of reducing the costs and ensuring the safety of the industry. It is also at the centre of politics because an array of players – from the government, the nuclear industry, the auxiliary industries, the scientific community, foreign affairs officials, to international nuclear vendors – compete to maximise their interests, whatever they may be.

In the intensifying debate about the merits of nuclear energy as a clean and sustainable energy source, scientists cannot even agree on the issue. Nuclear energy development, however, has to deal with the challenges of balancing the immediate consequences of burning fossil fuel and the long-term effects of radioactive waste management and the potential for nuclear annihilation. Who shapes the debate? What was the balance between the arguments over the immediate and pressing environmental pollution of burning fossil fuels and the longer-term issues of nuclear waste hazards? How have the changing social and economic conditions in the past 30 years in China affected this debate?

The plan of this book

No matter how small a proportion nuclear energy may contribute to the total electricity production in China, its development has much broader implication beyond the nuclear sector. This is especially the case when the nuclear industry is a truly global industry, any successes and failures in China have the potential to make the whole industry thrive or die. Furthermore, at the core of the nuclear energy development is the politics, which is not unique, but needs to be understood. Little has been written on the subject. Those who study China's energy policies seldom write about nuclear energy and those who write about nuclear power tend to focus on nuclear weapons programmes.[31] The literature on nuclear energy is primarily about developed countries where the influence of interest groups could make the industry possible but could also kill it. China remains a one-party-state that is not accountable to elections. Its political system and decision-making process are 'opaque and shrouded in secrecy' that require detailed examinations.[32]

This book is designed to fill the gap by examining the development of the nuclear energy programme in China over the past 30 years. The narrative of this development is designed not only to recount events, most of which have never been documented. More importantly, this account of events is designed to provide an understanding of the politics involved in nuclear energy development in China as a composite entity of Chinese politics. It will highlight the economic, technical, environmental and, more important, political challenges that nuclear energy development faces and the capacities that the nuclear industry needs to meet these challenges.

Chapters 2 and 3 provide this narrative in a sequential way to underscore four counter-intuitive stories about China's nuclear energy development: (a) the nuclear sector has always been torn by internal competition and rivalry and therefore for a long time no one in the sector wanted nuclear power projects; (b) the existing nuclear energy projects were not part of the energy development; rather they were initiated to achieve political, diplomatic or specific regional interests; (c) there has not been a close tie between the government and the nuclear industry, as usually exists in the industry around the world; and (d) the industry has enjoyed extensive access to information from abroad and close international cooperation. None of these characteristics are in line with the general thinking about Chinese politics – that is, the Communist Party has close control over a strategic sector, there is little distinction between the government and its state-owned corporations, and the Chinese would insist on their motto of 'self-reliance' in this strategic sector.

The four features that have defined nuclear energy development in China are grounded in their institutional bases. Institutions, as defined by Douglass C. North, have two main components – formal organisations that 'are consciously created and have an explicit purpose', such as government agencies, think tanks and other players; and informal ones, akin to the 'rules of the game', which refer to 'a set of common habits, norms, and interactions between individuals, groups and organisations.' Institutional structures and norms were crucial in shaping China's nuclear energy development.

Chapter 4 examines *the politics of nuclear energy development* in China by focusing on the changes in formal and informal institutions within which players interact. These players include different government agencies, regulatory bodies, state-owned corporations and increasingly 'non-state' players, such as think tanks and activist groups. In nuclear energy development, 'the lack of a cohesive, consistent national energy development policy is quite evident.'[33] It concludes there was no one 'decision-making point' where bargaining and compromises among different interests could even take place. Most nuclear power projects were initiated for reasons other than as part of an energy programme. Policy-making arenas shifted accordingly; sometimes they were predominantly foreign policy, at other times central-local relations. Nuclear power projects have become part of the energy considerations only in the past few years when demand for new nuclear power plants escalated and could no longer be managed on an *ad hoc* basis. It is yet to be proven whether the decision of 2008 to place nuclear energy under the auspice of a national energy authority, which itself may not have sufficient manpower or expertise on all energy issues, changes the fundamental rules of the game for nuclear energy.

The following four chapters then concentrate on specific challenges – economic feasibility, technological selection, nuclear fuel cycle capabilities and environmental considerations. Each of these four issues will be discussed in the context of the interaction among the formal organisational structure, key players and informal rules of the game.

Economic competitiveness – The cost of nuclear power is identified as the determining factor for nuclear futures. Many studies highlight the low economic competitiveness of nuclear energy as the result of its intensive initial investment, long construction periods and high risks. With rising fossil-fuel prices and worsening environmental pollution, however, 'the economics have moved in nuclear power's favour,'[34] argues the IEA.

The rapid rise of China as a major economic power within a time span of about 30 years is often described as one of the greatest economic success stories in modern times. From 1980 to 2008, China's economy grew 14-fold in real terms. China seems to be in a position to invest a great deal of financial and other resources into nuclear energy development. It does, however, remain a developing country where 'a group of relatively developed islands with a cumulative population of over 400 million people are scattered around in a sea of over 800 million people who live very much in developing-country conditions.'[35] Nuclear economics has to be discussed in this context.

There are several questions. First, how have the competing demands for financial capital and other resources been balanced? Which should have priority in receiving the financial support – nuclear energy or basic social and economic development? Second, given the country's intensive investment, a rush into nuclear energy would place tremendous pressure on the macroeconomic structure. How have the central and provincial governments, and different government agencies and large state-owned enterprises, cooperated in order to balance both economic stability and nuclear energy development? Third, who can invest, own, manage and operate nuclear power plants? On what terms are utilities going to be allowed to invest, own or operate nuclear power plants? Fourth, who should pay the cost (private or public, domestic or foreign) and under what conditions? How has the government established a pricing system that reflects the costs across the whole value chain and rewards good performance? At the core of these economic challenges is politics – which government institutions are in charge of investment and pricing and how are investment and pricing regulated?

Technology innovation – Technology selection and standardisation are the key factors in ensuring a sustainable and safe nuclear development. Although some argue that technological innovation is primarily driven by its internal logic and its inherent 'best' design, in practice technology selection and innovation are never immune from politics. In all countries that have developed a nuclear energy industry, there has been a consensus that one way to manage escalating plant costs and delayed construction times and to ensure safety of nuclear power plants is to standardise the technology. In all of them, there has been tension between those pressing for the industry's rapid development and those urging caution in technology development and selection,[36] and between those pursuing a 'technological nationalism' – 'national grandeur first and foremost in terms of technological prowess'[37] – and those willing to deemphasise national origins in favour of an internationalist vision and more rapid progress.

China has made an official decision to focus on pressurised water reactors, yet it has had different types of reactors built or under construction – from Russia, France, Canada, and the US and, of course, its own. What have been the debates in terms of technology selection and development in China? How China can standardise and localise imported technologies and develop its own brand remains a serious challenge for the government in its strategic policy making and for industry in ensuring both economic competitiveness and safe operation. Rapid expansion of any industry is a concern from the standpoint of assuring quality of construction and trained staff and operators. How to train craft labour, technical professionals and, more important, regulatory staff with technical expertise is a challenge for all countries that want to build a nuclear energy industry. These questions are examined in Chapter 6.

Nuclear fuel services – Concerns over energy security and surging fossil-fuel prices are among the important reasons for countries to choose nuclear energy. Though fuel constitutes a small portion of the total costs of nuclear energy and uranium reserves are much more widely available than oil and gas, securing an adequate uranium supply has been important for China. Chapter 7 examines the three-pronged approach China has adopted in securing its nuclear fuel supplies – expanding domestic uranium exploration, prospecting and mining, investing in overseas uranium exploration and mining, and increasing its imports of uranium, uranium products and fuel services. Each of these approaches presents challenges. In addition, China is developing an integrated nuclear fuel cycle with reprocessing capacity and fast breeder reactors to make nuclear fuel recyclable and the industry sustainable. China supports multilateral approaches to the nuclear fuel cycle. Yet, until recently, it had a bad record of nuclear export controls.

The challenge is how it can secure its position as a supplier while benefiting from multilateral fuel supplies when needed.

Environmental protection – Fossil-fuel combustion is a major source of worldwide environmental degradation, with coal the most polluting. Nowhere in the world do people feel the consequences more than in China. Nuclear energy is attractive because it is essentially free from CO_2 and other GHG emissions. Yet, China's leaders have been facing competing environmental concerns: indoor pollution as the result of lack of or low access to electricity, air pollution as the result of burning coal directly and the visual pollution of large nuclear power plants. There are also competing concerns about the immediate environmental pollution from fossil-fuel consumption and long-term radioactive waste management and decommissioning. Chapter 8 examines how the Chinese government has tried to balance these competing interests in its nuclear energy development, especially in the light of media liberalisation and the rising middle class.

Evidence shows that nuclear energy expansion in China is not, as is occasionally implied, devised out of a single plan from a dominant regime, transferred from the top level of government to its scientists and the industry. Decision making in this sector has always been based on a fragmented authoritarian structure. It has become more fragmented and competitive as the debate among various interests is brought into the open and has become increasingly lively. There is an argument about the relative contributions of coal and nuclear power to national development, between general economic and specific industry interests, between quick electricity profits and the development of local skills and ingenuity, between engineering and scientific communities, and between international and domestic perspectives and designs.

This public debate may be surprising in an arena that was traditionally regarded as intensely secretive and governmental. Nevertheless, the institutional fragmentation and new technologies have brought new players into the discussion. The issues under dispute, the different bureaucratic and scientific interests, and alternative strategies can all be discerned by careful reading of the evidence. There is no one dominant view, no unchallenged centre of knowledge and authority, and no guaranteed path of development. Nuclear energy policy making might have been liable to a burst of authoritarian directions from the top but without concentrated attention to its progress.

Without doubt, China will continue its nuclear energy development and expansion; nuclear energy is needed to address the twin challenges the country is facing – energy security and climate change. It would be unwise, however, to underestimate the tensions and debates and their impacts on nuclear energy development. The politics involved will continue to shape this industry and determine its future.

Notes

1. Evans, 2002.
2. Chubb, 1983; Haggard and McCubbins, 2001.
3. A NATO official describes the difficulties working as an international civil servant: 'to understand the daily life of an international civil service, consider the troubles of a national civil service, multiply by the number of member countries, and square it' with operation in both civilian and military fronts. An analogy can well summarise the challenges China faces – the large population base makes everything any government has to deal with exponentially more difficult, and low GDP per capita and low natural resources endowment further complicate the challenges. See Hans Mouritzen. 1990. *The International Civil Service: A Study of Bureaucracy: International Organisations*. Aldershot, UK: Dartmouth, p. xiii.
4. Ravallion and Chen 2004; IEA, 2007, p. 262; Yusuf and Saich, 2008.
5. Steinfeld, 2008, p. 133.
6. Kenneth Lieberthal, 'Statement to the Hearing before the Committee on Foreign Relations, US Senate', 4 June 2009, p. 7.
7. Ibid.
8. World Bank, 2007. Jonathan Schwartz, statement at the hearing on 'China's Energy Policies and Their Environmental Impacts', US-China Economic and Security Commission, 13 August 2008.
9. See Woo, 2007.
10. Price, 1990, p. 3.
11. Caro, 1982, p. 513.
12. Ibid, p. 528.
13. IEA, 2009, p. 96.
14. IEA, 2007, p. 349.
15. Zhang, 2007, p. 3547.
16. World Bank, 2008, p. 36.
17. Jiang Lin, Nan Zhou, Mark D. Levine and Favid Fridley. 2006. 'Achieving China's Target for Energy Intensity Reduction in 2010: An Exploration of Recent Trends and Possible Future Scenarios', Ernest Orlando Lawrence Berkeley National Laboratory, LBNL-61800, December.
18. World Bank, 2008.
19. Joanna Ingram Lewis, 'From Technology Transfer to Local Manufacturing: China's emergence in the global wind power industry', PhD thesis, University of California, Berkeley, 2005.
20. McElroy et al. recently, in the abstract of their article, claim 'wind could accommodate all of the demand for electricity projected for 2030, about twice current consumption.' In the text, they explain that 640GW of wind power installed capacity would be needed to accommodate a reduction in CO_2 emission of 30% of the additional capacity to be needed by 2030. This means (a) the remaining two-thirds of the additional generation capacity would have to be met by coal-fired capacity; and (b) the expansion of the current 12.2GW wind power capacity to 640GW in 20 years, that seems to be neither realistic nor likely in terms of physical location and operation or the investment required (see Michael B. McElroy, Xi Lu, Chris P. Nielsen and Yuxuan Wang. 2009. 'Potential for Wind-Generated Electricity in China', *Science*, 325, pp. 1378–80.
21. IEA, 2007, p. 274.
22. IEA, 2006b.
23. IAEA, 2006.

24. IEA, 2009, p. 647.
25. Feigenbaum, 2003; Lewis and Xue, 1988.
26. Falk, 1982; Camilleri, 1984; Byrne and Hoffman, 1996.
27. Price, 1990; Grimston and Beck, 2002; MIT, 2003, 2009; Joskow and Parsons, 2009; Young, 1998.
28. Campbell, 1988; Mounfield, 1991; Elliott, 2007.
29. Kursunoglu, Mintz and Perlmutter, eds. 1998.
30. Hans-Holger Rogner, 'Nuclear Power Revival: Short-Term Anomaly or Long-Term Trend'? World Nuclear Association Annual Symposium, 7–9 September 2005, London, p. 4.
31. Smil, 1988; Smil, 2004; Andrews-Speed, 2004.
32. Dumbaugh and Martin, 2009.
33. Lieberthal and Oksenberg, 1988, p. 25.
34. IEA, 2006b, p. 384.
35. Kenneth Lieberthal, 'Challenges and Opportunities for US-China Cooperation on Climate Change', Testimony before the US Senate Committee on Foreign Relations, (4 June 2009), p. 7.
36. Campbell, 1988, pp. 68–72.
37. Hecht, 1998, p. 65.

2
From Bomb to Power

By the late 1970s, China was the only nuclear power state without nuclear power; a state of affairs that caused a great deal of anxiety and embarrassment among the Chinese nuclear establishment. In the early 1970s, at an international conference on the peaceful use of nuclear energy, the chairman of the conference called the Chinese delegate to the podium, which had been reserved for those from countries that had nuclear power plants. The chairman said: 'We welcome the Chinese delegate to join us because, after all, Taiwan has built two nuclear power reactors.' The chairman's comments embarrassed the Chinese, who had to accept the invitation if they wanted to insist on a 'one-China' policy although they knew that at that time they did not have the ability to build nuclear power plants.[1]

This story may or may not be true, but it reflects the mentality of many Chinese policy makers and China's nuclear establishment of that era. China had begun its nuclear programme not much later than Britain and almost at the same time as France. Despite difficulties with international isolation, it had successfully pursued a nuclear weapons programme, building its first reactor in 1956 and testing its first atomic bomb in 1964 (with fission U^{235}) and its hydrogen bomb (thermonuclear device, fission-fusion-fission type using U^{235}, U^{238}, and heavy hydrogen) in 1967. Reprocessing and enrichment plants were also constructed in the 1960s, despite political turmoil and economic impoverishment.

China had done little, however, on a civil nuclear energy programme. From the 1950s to the 1970s, priority was given to the nuclear weapons programme in line with Mao's claim to break up the monopoly of nuclear weapons by a few Western states, and to demonstrate that China could do what other developed countries had already done.[2] Nonetheless, a blueprint was approved in 1958 to develop technology that would produce and control the release of energy from splitting the atom, a technology that could be used for nuclear submarines and civilian nuclear power plants.

The dual nature of the programme led to constant competition for attention and resources between the civilian and military branches of the nuclear establishment. The reactor programme was abandoned in 1962 because of a severe shortage of financial, material and human resources. A group of scientists working on the programme were moved between various institutions. First, they were removed from the 2nd Ministry of Machine Building, which was in charge of nuclear development, and placed under the 7th Academy of the Commission of Science, Technology and Industry for National Defence (COSTIND); then, in 1965, they were sent back to the 2nd Ministry.[3] Without political patrons or a stable institutional base, the reactor programme existed in name only.

In the 1970s, a nuclear energy industry was blossoming in the West, but in China resources went into the weapons programme while electricity shortages haunted all sectors. Then Premier Zhou Enlai (周恩来) ordered that a portion of nuclear research and development be shifted from military to civilian use. A decade later, the State Council approved two nuclear power plant projects: the Qinshan (秦山) nuclear power plant (NPP) in November 1981 and the Daya Bay (大亚湾) NPP in December 1982. The two projects represented two trends of nuclear energy development – one relied on self-finance and indigenous technology and the other mirrored the reform that was underway, with new thinking, new ways of financing and new ways of managing a large project. Different sets of politics were involved. The development of these two projects has also set the trend for nuclear development in China, which is rife with contention, competition and rivalry.

The issues debated and fought over included: (a) whether China should build NPPs when it was so poor; (b) what priority the projects should be given when there was abundant cheap coal and large hydro potential; (c) how the projects could be financed; (d) whose technology should be used for the nuclear projects; (e) where nuclear power plants should be located; and (f) who should lead the efforts. In the late 1970s and 1980s, the division over these issues was sometimes between the Ministry of Nuclear Industry (MNI) and the Ministry of Electric Power (MEP), sometimes between the central and provincial governments, sometimes between the Ministry of Finance and line ministries, and often among various competing interests. Nuclear development in China, therefore, is an exemplar for understanding politics in China in general and reform in particular.

The bomb programme

China started its nuclear programme in the 1950s. Although Mao had once called nuclear weapons 'paper tigers', he demanded a nuclear programme to improve China's international status and its military power. 'As for the atomic bomb, this big thing,' Mao was quoted as saying, 'without it people say you don't count for much. Fine, then we should build some.'[4]

Geologists were asked to look for the country's uranium reserves as early as 1953–54, while nuclear physicists were directed to start researching on heavy water reactors and accelerators. In the 1950s, the country had few technical experts, except for a small number of scientists who had been primarily educated in the West: mainly in Britain, France, Germany and the US. They filled the positions in the newly created Institute of Physics, headed by Qian Sanqiang (钱三强), a nuclear scientist trained and worked at Institut Curie in France. In a bilateral agreement signed in 1957, the Soviet Union agreed to help China explore uranium mining, develop a nuclear weapons programme and train scientists at Soviet universities and laboratories. A Joint Institute for Nuclear Research was established in Dubna, in the eastern part of the Soviet Union, where more than 1000 Chinese scientists were eventually trained. A team was assembled in China. At least 60% of them were under the age of 22 and few had university degrees. Resources were poured in at a time when millions of people were close to starvation and many promising students were then sent to the Soviet Union for their undergraduate and postgraduate studies in the following decade.[5]

In 1958, to support the efforts, the Chinese government merged the then Scientific Planning Commission and the State Technological Commission into a State Science and Technology Commission (SSTC) to oversee the civilian side of science and technology research, including nuclear science. It also created the Commission of Science, Technology and Industry for National Defence (COSTIND), headed by Nie Rongzhen (聂荣臻), a general in the People's Liberation Army (PLA). COSTIND had dual accountability to the Central Military Commission (CMC) of the Chinese Communist Party (CCP) and the State Council. This team led by Nie Rongzhen would 'control the scientific and technical resources of the PLA, the State Council's military industrial system, and the defence-related sciences of the Chinese Academy of Sciences.'[6]

Two other major reorganisations took place in the 1950s: the 3rd Ministry of Machine Building was renamed as the 2nd Ministry of Machine Building (hereafter called the 2nd Ministry), which was responsible for the nuclear development programme (civil and military), and the Institute of Modern Physics of the Academy of Sciences, which was renamed as the Atomic Energy Research Institute (AERI) in 1958 and placed under the 2nd Ministry. It was directly accountable to COSTIND and thereby the CMC. In the decades that followed, the 2nd Ministry was the centre for nuclear research and its application in China and the AERI was where most nuclear scientists and experts were located. The significance of relocating the AERI from the Academy of Sciences to the 2nd Ministry (and thereby COSTIND) was that these experts were placed under tight military and political control. This meant that any competition for resources and control would involve internal politics of the 2nd Ministry.

The arrangement was also a way to insulate the scientists and experts from political campaigns launched by Mao and protect them from political persecution. This was at a time when the anti-rightist movement particularly targeted intellectuals who were not interested in politics or what the party was doing. Indeed, according to some sources, General Nie 'quietly transferred most of the civilian employees at the nuclear and missile research institutes into military service' to protect them from being investigated and persecuted.[7] Many more, however, were relocated to the newly created field in the Gobi Desert, far away from the political epicentre and to a place where the atomic bomb tests would eventually take place.

Just after the organisational structure was put in place, and while scientists began working on their first blueprint of the atomic bomb, the Soviets told the Chinese government on 20 June 1959 that 'because of negotiations on a test ban under way in Geneva, Moscow would not supply the prototype bomb or blueprint and technical data on the bomb.'[8] With their relationship with China deteriorating rapidly, on 23 August 1960 the Soviet Union withdrew all its 233 experts working on nuclear programmes in China. Although China was suffering from a terrible famine in the aftermath of the Great Leap Forward and everything was in short supply, Mao decided that the bomb programme would continue.

The weapons programme continued at the expense of many other programmes, including the one on nuclear reactors. In 1962, the Politburo decided to form a Special Commission led by Premier Zhou Enlai, and consisting of the premier, seven vice-premiers of the State Council and seven ministers, to be in charge of the nuclear weapons programme. It also decided to close down some other programmes, including the one on nuclear reactors, so that resources could be poured into the bomb programmes. To save the experts and their research, some proposed to place them under the control of the Navy, which was unable and unwilling to host them. In 1962, this group of scientists (fewer than 50) was moved out of the 2nd Ministry and placed under the Institute of Atomic Energy under the COSTIND. After the first bomb test in 1965, the Politburo decided to revive the reactor programme as part of the nuclear submarine development and ordered the 2nd Ministry to complete the test model of a nuclear submarine by 1970. The 2nd Ministry brought back from COSTIND a small team of experts, researching nuclear reactors, to work with the experts on nuclear weapons programmes. Less than a year later, the Cultural Revolution started and political turmoil affected everyone involved in the project. In 1969, some scientists working on the nuclear submarine programme were placed under the Navy to shield them from political persecution. Just before the first nuclear-powered submarine was sent to sea for trial-testing in 1971, the Navy made a decision to move the whole team back to the 2nd Ministry.[9]

In a little over a decade, the research on nuclear reactors went through seven rounds of reorganisation:

1958	Emerged as the nuclear energy division within the Institute of Atomic Energy, which was moved from the Academy of Sciences to the 2nd Ministry
1960	Merged into the Design Bureau of the 2nd Ministry
1962	Moved to the COSTIND
1965	Moved back to 2nd Ministry
1966	Moved to the COSTIND
1969	Moved to the Navy
1971	Moved back to the 2nd Ministry

In contrast to the nuclear weapons programme that was under a centralised control led by General Nie and located far away from the eye of the political hurricane (in Beijing), the nuclear reactor programme went through constant organisation and reorganisation, changing leaders, changing personnel and changing locations,[10] and then they were subject to the frequent political turmoil that was sweeping the country. Some who were involved in the project for decades argued that the short-sighted view of leaders was the main reason for the failure of China to have developed its own technology of nuclear reactors.[11]

In sum, several important developments in the bomb programmes have had deep impacts on nuclear energy development: (a) the nuclear weapons programme had always received top priority on talent and resources; (b) nuclear development was subject to political campaigns launched by Mao as all the other sectors in the country; (c) nuclear development was affected by political and economic isolation and (d) both military and civilian aspects of nuclear development had to deal with bureaucratic fragmentation and competition. Meanwhile, two developments were critical for the nuclear energy programme once the reform started in the late 1970s: (a) the nuclear weapons programme kept a team of scientists and experts alive and active, despite constant political campaigns. This included an older generation who were trained in the West and returned to China in the early 1950s, a group of scientists trained in the Soviet Union, and those trained in Chinese universities. By the late 1970s, there were about 4000 of them, who had suffered political persecution but not as badly as many intellectuals outside the military. And (b) the weapons programme allowed scientists to continue their research and experiment on reactor and related programmes, all of which made a switch to a civil nuclear programme possible.

Moving into power

An interest in harnessing the dual-use potential of nuclear technology did not rise in China until the 1970s when the country faced a severe shortage of electricity supply. While nuclear energy in general seemed to be accepted by Chinese political leaders, doubts and debates about whether the country should develop a nuclear energy programme, how it could do it and who would be in charge were prevalent. The internal political turmoil made all this discussion moot and the actual programme was put in place only after the economic reform started in 1979.

When the Cultural Revolution came to an end, a generation of 'old' revolutionary cadres with reforms in mind, such as Deng Xiaoping, returned to power. They were willing to try new ideas and even to speak up, at least within the decision-making circle. It was then accepted that: (a) it might not be a bad idea to start a nuclear energy programme, despite that the country had many other energy resources, especially coal and hydropower, and (b) there might be a way to get sufficient financial resources to start the programme if they were willing to think beyond the traditional framework. Resistance was real too, from the political leaders, the nuclear establishment and ordinary workers in the related industries. By the early 1980s, a nuclear energy programme was finally put in place and details were being worked out for the two quite different nuclear power projects. The debate over the first two NPP projects and the process to make them a reality were significant because they mirrored the vicissitudes in the early stage of the reform.

Initiatives

In early 1970, just before the Chinese New Year, an emergency report was sent to the central government stating that a severe power shortage had forced many factories in Shanghai to close. At the time, industry in Shanghai accounted for one-sixth of the national total industrial production. The immediate cause of the shortage was that the city was running out of coal. Transporting coal to generate electricity took more than 70% of the rail capacity in China. Political turmoil and a bottleneck in transport infrastructure paralysed the industry in Shanghai.

On 8 February 1970, at a State Council meeting, Premier Zhou Enlai outlined his view: because the coastal regions were short in energy resources, in the long run they needed to develop nuclear generation capacity to solve the problems of electricity shortage. Zhou told the 2nd Ministry that it should not only focus on the weapons programme but also begin researching on nuclear energy. Zhou's speech signalled the beginning of China's nuclear programme, known as the '728 project', which represented the date of the speech.

The municipal government in Shanghai and the Chinese Academy of Sciences Shanghai branch jointly assembled a team of scientists from universities and research institutes. Ouyang Yu (欧阳予), a Soviet-trained nuclear scientist, was brought back from a labour camp to lead the team as chief engineer for the 728 project. The 2nd Ministry sent eight experts to join the team. It took more than two years to decide who was qualified to work on the project and about 4000 people were transferred from military to work on it with the team.[12] The team was placed under the Shanghai Institute of Nuclear Research, also known as the '728 Institute', which was transferred from the 2nd Ministry to the Shanghai government in July 1974.

As soon as the team was assembled, fights erupted. Those from the nuclear establishment complained that civilian engineers and scientists were neither trained in the field nor qualified to work on a nuclear energy project and that they did not even know how to draw a blueprint. Those from universities and research institutions said that the nuclear establishment might be able to build a bomb, but it could not handle more sophisticated nuclear power technology. Some wanted the research to focus on pressurised water reactors (PWR) that had been used by more than two-thirds of the world's nuclear power stations. Those from the 2nd Ministry who had worked on experimental heavy water reactors (HWR) insisted that the new project be based on that technology.

At that time, China had only two small research reactors in Beijing: a swimming-pool type unit of 3.5MW and a 7MW heavy-water-reactor. Both had been developed as part of the weapons programme under the 2nd Ministry. The bitter argument that ensued was not only about who was in charge but also about the direction of China's nuclear programme and the way it would develop. More importantly, it reflected a power struggle among different factions.[13]

The Cultural Revolution was in its death throes and political rivalries made it impossible for the project to proceed. The 728 project was quickly hijacked by the Gang of Four, which had its stronghold in Shanghai. Given that the 728 project was Zhou's initiative, the Gang of Four could not reject it outright. Instead, they demanded the country be rid of 'capitalist intellectuals' in defence industries while calling on it to 'catch up with the West in nuclear development in three years and then surpass it in the following two years'. Ironically, this achievement would depend on the very people they had tried to remove.

The radicals insisted that China must develop something so advanced that no other country would be able to match them in such a short period. Against the advice of scientists, they ordered the 728 Institute to put all the resources in a project to develop a molten salt reactor (MSR). The concept of MSR emerged in the 1960s, but no one had believed that the technology could reach an application stage in the near future. Indeed, in 2002 at an international forum called by the American government and attended by the government-supported research institutes in a few countries (UK, Switzerland, South Africa, Japan, France, Canada, Brazil and China) to discuss the fourth generation of nuclear technology, scientists decided that six concepts could be used to develop the fourth generation of nuclear reactor and one of them was MSR. The Gang of Four insisted that the Chinese scientists could complete the research of MSR in three months and refused to put resources elsewhere.

More than four years later, and after pouring in more than a million yuan (huge resources considering that the country was constantly short of capital), no progress was made. The chief engineer of the 728 project and another top nuclear scientist who had led the research on reactors for nuclear submarines took the issue back to Zhou Enlai. In March 1974, Zhou was already ill and at the meeting of the Special Commission he chaired for the last time, the issue of technology for the 728 project was raised. Regarding MSR, Zhou asked whether the technology was safe and whether it had been used in the US. After being told that the technology was still under-researched and would 'probably' be safe, he rejected the idea outright. Zhou instructed that nuclear energy development must follow four principles: safety, economy, practicality and self-reliance,[14] and the Shanghai 728 project would use the PRW technology, as suggested by scientists, based on its nuclear submarine experience, with a capacity of 300MW. In April of that year, the State Planning Commission (SPC) included it in the national economic plan for 1975.

Four modernisations

One guiding principle behind the 728 project initiatives was the 'four modernisations'. In January 1975, in his last public appearance outside hospital, Zhou delivered his government report as the premier at the 4th National People's Congress. In the second stage of his two-stage development plan (1980–2000), Zhou stated that the country would pursue 'the comprehensive modernisation in agriculture, industry, defence, and science and technology'.

By February 1975, Zhou was too ill to be in charge and Deng, who had been brought back from the cold in 1973 by Zhou, was put in charge of the daily work of the State Council. Throughout 1975, in a series of documents and speeches, Deng argued that the priority of the country should be to improve productivity rather than concentrate on 'class struggle', and the way to achieve the four modernisations was to restructure the economy by opening up to foreign trade and providing more incentives to local governments and individuals.[15] The four modernisations, along with six suggestions on reforming the economy, were quickly identified by the Gang of Four as the 'three poisonous weeds', and Deng became the target of the 'anti-rightist deviationist wind campaign' while he was still in charge of the State Council. It took another year, a critical period of power struggle between the Gang of Four and the moderates, before the four modernisations were formally adopted as official policies in 1977.[16]

In August 1977, at the 11th National Congress of the CCP, in announcing the end of the Cultural Revolution, Deng also restated the four modernisations, which in practical terms meant: electricity in the rural areas, industrial automation, a new economic outlook and greatly enhanced defence strength. In March 1978, at the National Science Conference in Beijing,

Fang Yi, vice-premier in charge of the State Science and Technology Commission, listed nuclear energy, high-energy research and nuclear fusion research as Chinese scientific and technological priorities. Fang Yi also provided a more detailed plan of action to achieve the four modernisations at the conference. The plan included:

- Agricultural technology.
- Energy – nuclear power, solar energy, etc.
- Natural resources – iron ore, copper, aluminium, nickel, cobalt, titanium, vanadium, etc.
- Computer sciences – giant computers, computer networks and serial production.
- Aerospace technology – satellites, skylabs and space probes.
- High energy physics – proton accelerator development.
- Genetic engineering – molecular genetics and biology.
- Anti-pollution technology.

Fang Yi commented: 'Atomic power generation is developing rapidly in the world, and we should accelerate our scientific and technical research in this field and speed up the building of atomic power plants.'[17] He also emphasised the need to introduce foreign technology. An important component of this plan was an exchange of scientists between China and other countries, especially the US. Chinese Academy of Sciences and the US National Science Foundation (NSF) signed an agreement for exchanging students and scholars in early 1978. In 1977 and 1978, the Institute of Atomic Energy in Beijing also entertained delegates from France, Germany and Japan and in return, sent its nuclear scientists to these countries. As one of the first groups of scientists invited to the US, Chinese nuclear physicists visited Illinois National Accelerator Laboratory in 1978. They told their American colleagues that China was interested in purchasing American technology, including two or three nuclear reactors by the end of the decade.[18]

NSF sent a mission of academic and commercial representatives to Beijing in the same year. At the request of their hosts, the Chinese Academy of Sciences, the Americans gave lectures in their specialties: pressurised water reactors (manufactured by Combustion Engineering), radioactive waste management, metallurgy and fuels, education of nuclear scientists and engineers, high-temperature gas-cooled reactors and research reactors based on Triga-type reactors (manufactured by General Atomic). The papers on pressurised water reactors and high-temperature gas-cooled reactors were specifically asked for in advance.[19]

These activities gave the world an indication that China was serious about nuclear energy development. The Executive Director of the American Nuclear Society, Octave Du Temple, stated after a visit to China in April–May

1978, that: 'It is clear that China will acquire probably two or three nuclear reactors from the West within the current eight-year plan.'[20]

Power shortages

Throughout the 1970s and 1980s, power shortages were one of the main obstacles to economic growth. It was estimated that 25–40 % of China's productive capacity lay idle at the end of the 1970s as the result of power shortages. Coastal regions suffered even more. In Guangdong, for example, only 61% of the electricity requirements of its industries could be met in 1980. Consequently, the province endured 7500 million yuan losses in industrial output in 1979.[21] Electricity shortages led to a series of editorial articles in the official newspapers, *People's Daily*, *Economic Daily* and *Guangming Daily*. Some identified this shortage as the result of lack of generation capacity. Electricity generation in China accounted for a mere 2.8% of the world's total power generation, while the country had 20% of the world's population. Electricity consumption per capita in China was 247 kilowatt-hour (kwh) in 1978 in comparison with the world's average of 1,527 kwh and 9,604 kwh in the US. Many emphasised that the problem was not shortage but waste and inefficiency. Energy intensity in China (energy used per unit of GDP) was more than 50% above the average of advanced countries. This poor performance was largely due to very low efficiencies in coal combustion. According to the minister of Electric Power, 'Average 1978 coal consumption by large thermal power plants under the ministry was 433 grams of standard coal per kilowatt-hour, [compared] with the current US average of about 360 and the Soviet mean of 330 grams of standard coal per kilowatt-hour.'[22] 'Much of our energy is wasted with the country's inefficient 180 000 boilers', proclaimed one newspaper editorial, continuing 'If we can replace them with larger and more advanced boilers, we can save more than 20% of the 200 million tonnes of coal consumed each year'.

Whether the root of the problem was the lack of generation capacity or inefficient generation capacity, power shortages impeded economic reform. How to solve the problem of power shortages, however, was a topic for debate. Some saw the solution as an increase in coal production and expansion of thermal power generation capacity. In 1978, Liu Lanbo (刘澜波), minister of Electric Power, announced through the *People's Daily* that 'The general line of China's energy policy is to develop hydropower and coal resources, in accordance with local conditions.'[23] Opponents argued that coal might be abundant in China, but not along the coastal regions and transporting coal to power stations in these places became particularly problematic because of the bottlenecks in the rail system. Some paid attention to developing renewable capacities – small hydro stations, biomass, photovoltaic arrays or wind turbines that would operate independently from electrical grids, but they also realised their limitations.[24] Oil, though fuelling many of the country's power stations at the time, was regarded as 'too precious a commodity to burn.'[25]

Energy problems, particularly shortages of electricity, became one of the key issues discussed at a meeting of the CCP Central Committee's Leading Group on Finance and Economics held on 21 June 1980. After heated discussion, the participants agreed that the country would increase coal production by 200 million tonnes during the 6th FYP (1981–85) and add 4GW generation capacity each year in the next five years. They also agreed to lower the entry barriers: local and even foreign investment would also be allowed in coal and power generation. Finally, they decided the country would increase its oil exports in order to get foreign exchange to support the reform. Li Peng (李鹏), then deputy minister of Electric Power, argued that some of the revenue from oil exports should be allocated to the power sector development. When the committee decided to allocate 10% of the revenue to the power sector, he argued for more but failed to get what he wanted. The participants also decided that, to the outside world, the official policy was 'the combination of energy development and energy conservation' with no specific sector given priority. Regarding nuclear energy, the committee simply passed on Deng's message that China would build two nuclear power stations somewhere along the coast. However, no specific decisions were made. Nuclear energy sounded a great idea, but few knew how China could develop it when the country had neither the resources, nor human capacity, nor technology.[26]

Facing the energy shortages, some in the nuclear establishment responded immediately. Given the bottleneck of coal and transport, unstable petroleum prices and supplies, depletion of fossil-fuel resources and the country's underdeveloped and fragmented grid system, nuclear energy was championed as a real alternative energy source in China's pursuit of the four modernisations. It was argued, nuclear technology was mature, and nuclear energy was safe, reliable and economically viable. It became the only new energy resource capable of alleviating the energy shortage on an industrial scale.[27]

The nuclear establishment argued: (a) a large-scale introduction of NPPs in the densely populated industrial areas would help overcome the transport bottleneck associated with the further expansion of coal output; (b) with the introduction of nuclear power, electricity generation would become less dependent on coal production and (c) development of nuclear power stations would help deal with the problems of pollution from coal-fired plants, of which the Chinese were increasingly aware – existing coal plants had no scrubbers or electro-static and produced the most pollution in the country, including, ashes, particles and CO_2.[28] 'China cannot wait any more to build a couple of nuclear power plants, as France, Japan and the US did, to alleviate power shortages and reduce some of the pollution as the result of coal burning,' stated one editorial article in the *People's Daily* in 1980

Scientists presented their arguments to policy makers when they were invited to give seminars to the members of CCP Central Committee. They were also brought in to brief members of the Politburo, which included Hua Guofeng, Li Xiannian, Yu Qiuli, Yao Yilin, Hu Yaobang and Wan Li.[29] These were the politicians who held the real decision-making power at the time. Such consultation could only take place because of the initiation of the reform. After nearly a decade of complete institutional paralysis and the destruction of education during the Cultural Revolution, in 1977 Deng led the decision to reopen universities as one of the new approaches towards the relationship between Marxism and the natural sciences on the one hand and between science and technology and economic development on the other.[30] Science and technology were seen as the motors of the future growth and modernisation of the Chinese economy. Bringing scientists, especially the leading ones, back from labour camps and giving them the opportunity to resume some of the research became part of the effort to fit the needs of readjustment and economic reform. Very quickly, these scientists became the driving force behind 'China's nuclear lobby... pushing for nuclear energy out of self-interest.'[31]

Jiang Shengjie, a well-known nuclear scientist and the president of the Chinese Nuclear Society wrote in 1984: 'At the present rate of coal extraction, China cannot keep up with electric power requirements... [W]ithout nuclear power China would find it impossible to achieve its programme of industrial, military and agricultural development by the year 2000 target date.'[32] While the nuclear establishment might have presented 'a firm, one might even say, blind commitment to nuclear energy,'[33] those in charge of electricity in the Ministry of Water Resources and Electric Power (MWREP) had quite a different view.

Indeed, with its 'two quite different types of agencies' having contacts 'with different portions of the Chinese bureaucracy,'[34] the MWREP held several positions on the nuclear issue. Agencies representing water wanted to see resources going into the development of hydro stations while some sections of the power agencies wanted to see the development of thermal power plants. A few wanted a rapid expansion of nuclear energy but were impatient with the domestically developed technology to meet the rising demand. According to one observer, two leading officials of the MWREP were publicly opposed to nuclear power because, they argued, it would take too long and too much investment to build a nuclear power plant with the same capacity of a thermal power plant, which would take approximately 18–20 months and a fraction of the capital to complete. The debates carried on in the first half of the 1980s as different segments of the energy industry were fighting for resources.

Taking-off

On 4 December 1978, at a press conference on a Sino-Franco economic cooperation agreement, Deng Xiaoping announced that China had already decided to purchase two nuclear power stations from France. The announcement surprised everyone. It might be true that from 1978 to 1993, 'no major policies were adopted of which Deng did not approve, and Deng himself was the initiator of many important policies.'[35] Deng neither managed economic policy on a daily basis nor intervened in economic policy-making on a regular basis. As a 'hands-off leader', Deng established a general orientation for policy while other policy makers, especially those in charge of the macro-economy and those at line ministries, had to come up with specific plans to bring his announcement to fruition.

It became clear from that point on that the debate was no longer whether China would start a nuclear energy programme, but on: (a) how to finance it; (b) whose technology would be used; (c) where the first nuclear power station would be built and (d) who would be in charge. 'The Chinese energy industry is part of the complex, hierarchical, Chinese political-economic system,' wrote Lieberthal and Oksenberg.[36] In this system, 'to translate their policy pronouncements into reality, the top leaders must use their limited leverage and weave their policies and projects into the existing web of bureaucratic exchanges through a protracted process of negotiations and consensus building.'[37] What is interesting for this study is not this bargaining process *per se*, but that debates, bargaining and lobbying actually took place over one of the most sensitive issues at a time when the country was just starting to open up.

Several key institutions involved in the decision over the nuclear energy development included the State Council, State Planning Commission (SPC), State Economic Commission (SEC), Commission of Science, Technology and Industry for National Defence (COSTIND), MWREP, Ministry of Nuclear Industry (MNI, the successor of the 2nd Ministry of Machine Building), Ministry of Urban and Rural Construction and Environmental Protection, Ministry of Metallurgical Industry, Ministry of Geology and Mineral Resources, Ministry of Finance and Ministry of Foreign Affairs. 'Regional interests representing such areas as Guangdong and Liaoning, which [had] their uranium, also [saw] the nuclear option as a way to reduce dependency on outside bureaucracies that controlled fuel and transportation.'[38] Two projects were at the centre of the debate – the 728 project, later known as the Qinshan I, and the Guangdong project, later known as the Daya Bay project. Each involved its distinct politics in their preparatory stage: the 728 was fought over among the nuclear establishment while the Daya Bay project was pushed by the provincial government in Guangdong.

The 728 project

In February 1978, Li Xiannian (李先念), the vice-premier in charge of the economy, approved a report on the 728 NPP project submitted jointly by the SPC, the State Construction Commission (SCC) and the COSTIND. The State Council also decided to move the 728 Institute out of the hands of the

Shanghai Municipal Government and placed it back to the 2nd Ministry. Following the 3rd Plenum Session of the 11th Central Committee of the CCP in December 1978, where the decision was made on reforms, it was time to decide again on how to push the 728 project ahead.

On 31 January 1979, Gu Mu (谷牧), vice-premier of the State Council in charge of the economy, and the secretary of the Party Central Committee Secretariat, also heading the Foreign Investment Control Commission and Imports and Exports Control Commission, called a meeting on behalf of the State Council to discuss the 728 project. The meeting was attended by ministers from SPC, COSTIND, the State Construction Commission, the 1st and the 2nd Ministry of Machine Building and MWREP.

Three of the six ministers supported the continuation of the project and three opposed it. Gu Mu chaired the meeting and had to cast his deciding vote. He stated that the 728 project should continue for the foreseeable future before the country could obtain the foreign technology for nuclear power stations, because the government had already invested so much into it. Meanwhile, he also qualified his decision by saying that the project, nonetheless, should not be taken as a starting point for the systematic development of a nuclear energy industry in China: the 728 project was an experimental project for the scientists to learn all the necessary technology for potential future development.

The Ministry of Electric Power paid little attention to the 728 project. As far as those in the power sector were concerned, the 728 project was neither an energy project, nor a commercial project. It was as before, a defence project under the jurisdiction of the 2nd Ministry, and its funding would be from the defence budget, not from the budget for energy development. Oddly, the 2nd Ministry did not take the project seriously either because, as far as it was concerned, the 728 project was an energy project. The Ministry insisted only on its control over nuclear fuel production and utilisation but not the rest of the project. Neither ministry wanted to take the lead, especially in ensuring its funding. This vacuum caused many concerns among those nuclear scientists who had argued for nuclear energy development.

In 1980, 'some 100 nuclear energy scientists and specialists advised the government that nuclear energy development should be made a long-term stable policy for meeting the country's energy requirements.[39] 'In February 1980, a barrage of arguments in favour of nuclear power poured forth at the first congress of the Chinese Nuclear Society, again at the second congress of the Chinese Scientific and Technological Association in March, and in featured articles in the *People's Daily*, and *Guangming Daily*.'[40] They proposed to construct six nuclear power stations in China, with two each in Guangdong province, in East China and in Liaoning province, all these places of acute power shortage. Some leading scientists, who had also recently taken on high official positions, such as Jiang Shengjie, vice minister of the 2nd Ministry, and Peng Shilu, chief engineer for the nuclear submarine programme, made their views known by giving lectures to the Party and government officials and writing featured articles in the official newspapers, such as *People's Daily* and *Guangming Daily*. A few leading nuclear scientists, including Qian Sanqiang and Wang Ganchang, were asked to give lectures to the top officials of the Community Party. The reasons for nuclear energy programmes included:

(a) Developing nuclear energy on a large scale was a fundamental measure to help the country meet the rising energy demands.
(b) Developing nuclear energy in industrial and densely populated regions would help alleviate the bottleneck pressure from the coal and transport sector and the limitation of the country's power grid system.
(c) China had already completed scientific research and experiments on nuclear power for military use and it would be easily adopted for civilian nuclear programmes.
(d) China had sufficient uranium ore to operate nuclear power plants in addition to satisfying military use.
(e) Nuclear energy was as economical as thermal energy, with little pollution.[41]

These scientists had the support of Deng in particular, who was willing and ready to open the country to the outside world as fast as possible. They also had the support of a few in the military. China had already invested heavily in the nuclear weapons programme and it needed to reap some urgently needed economic benefits in the form of power generation. 'By the early 1980s, mainland China had a large and diverse network of nuclear research institutes, more than ten experimental and production reactors, uranium enrichment plants and other facilities' and the estimated employment in this sector was between 100 000 and 300 000.[42] 'If we do not make an early decision about the principle of development for nuclear energy and allow things to be put off, this would not only be detrimental to the development of the nuclear industry but will also bring about waste and loss among the nuclear power science and technology forces,' a *People's Daily* article stated in June 1980. An official from the 2nd Ministry wrote an article in *Guangming Daily* in December 1979, sending out a similar message: 'We must see that if the human potential is not brought into play, the existing contingent will be lost. Such losses will be irreparable.'[43]

The scientists' proposal might have been supported by the policy makers in principle, but without bureaucratic endorsement from ministries or political patrons from the provinces scientists could not have fought the battle over resources allocation. The initial decision to base the 728 project in Shanghai was made partly because the city had research facilities and partly because of the uneven geographic allocation of resources. Coal was mainly in North and Northwest China while Southwest China was rich in hydro resources. Therefore, it had long been decided that if the nuclear project was to go ahead it would have to be located along the east coast. Anhui, Jiangsu, Shanghai, Zhejiang and Guangdong had been considered.

The government in Shanghai did not want to host a nuclear power station because of its high population density. Jiangsu was worried that the waste water from a nuclear power plant might harm its agriculture, while its southern neighbour, Zhejiang, did not want it because of the concerns with its fishery industry. Without 'both vertical and horizontal' support from the top leaders of the ministries, commissions and provinces, the 728 project did not have the political patrons to exert significant pressure and influence in the decision-making process.[44] On 22 April 1981, Gu Mu was recorded saying that the central government still had to decide whether the 300MW project led by the 2nd Ministry would go ahead. Two months later, Gu Mu instructed the 2nd Ministry and the Ministry of Electric Power (MEP) to form a leading group to take charge of nuclear energy development in China.

On 31 October 1981, the State Council finally approved the feasibility report on the 728 NPP project, located in Qinshan, Haiyan County, Zhejiang province, along the coast about 50 km from the Shanghai border. In Qinshan, the population was sparse; its seismic conditions were favourable and there was enough space for the installation of two 300MW units. The State Council reportedly allocated $100 million in foreign exchange to the project. The project was supposed to be domestically designed and domestically built, even though it was acknowledged from the beginning that 'a few key components – reactor coolant circulating pumps, several lesser pumps, and the neutron flux-mapping system, as well as a few complex castings and forgings' would be imported.[45] Another round of meetings was held in April and May 1982 between the 2nd Ministry and MEP that focussed on several issues: safety, technology and financing. It was agreed that even after an accident at the Three Mile Island nuclear generating station in the US, it was safe to build a nuclear power station with the current technology. A 300MW pressurised water reactor might be smaller than would be considered economical in the West, but the Chinese felt it was more manageable for the first try. The project would be constructed predominantly with its own finance and the foreign exchange would come from coal exports in Jiangsu and Anhui.

On 2 November 1982, the project was formally renamed as the Qinshan project and the 728 Institute changed formally to Shanghai Nuclear Engineering Research and Design Institute (SNERDI). SNERDI was given the responsibility of designing the reactors and the Electrical Energy Institute of Eastern China, the Shanghai Turbine Plant (under the Ministry of Machine Building) were assigned to design and manufacture the turbine. Zhejiang province made sure that water from the plant would be channelled to the sea so that it would not affect its fishing industry. The relatively small size of this NPP 'underscores the fact that the economics of power planning is not the primary consideration behind 728.'[46] The principle of the Qinshan project was shown in a slogan that had been clipped into Chinese characters in bushes outside the test centre: *self-reliance and hard work* (自力更生, 艰苦奋斗). On 1 June 1983, the project formally started.

Much of the debate over the Qinshan project was political and the decision to implement the project was also made for political reasons. China needed to develop nuclear energy not only to generate electricity but also to demonstrate its ability to do so as a nuclear power and to keep the contingent of nuclear scientists and engineers together and alive at a time of rapid political and economic changes. Given these broad political issues, the project did not have specific political patrons fighting for its interests, except for a few nuclear scientists. None of the three key government agencies contributed many resources to the project: the Ministry of Nuclear Industry (a successor to the 2nd Ministry) insisted on its control over nuclear fuel production and distribution, but was just beginning to find its way into the nuclear energy field. It was 'manifestly clear that the authorities, in particular those of the MWREP and of the Ministry of Machine Building, had no intention whatever of patiently awaiting the coming about of the development.'[47]

Daya Bay project

The nuclear project in Guangdong was a completely different story. It was not only a nuclear project but also a battlefield where the central and provincial government, the nuclear and electricity ministry, reformers and conservatives, and the Chinese and foreign governments were fighting for control, influence and interests.

Initiative

If the main controversy over the Qinshan project (also known as the 728 project) was whether the country should start a nuclear power programme, the Daya Bay project was started simultaneously with completely different objectives in mind. While the Qinshan project almost suffered an early death, the Daya Bay project emerged with hope for change. Its proponents, especially those in Guangdong province, had a much broader objective in mind than simply building a power generation plant. Therefore, the issues involved in the Daya Bay project were not whether the project would fly, but how it could take off, where it would be located, how it would be financed and who would be in charge.

A large body of literature is available on the origins of economic reform in China. What is important for this study is the rise of Deng and the provincial government in Guangdong. 'Without such a coalition, Guangdong's take-off would not have been possible.'[48] When Deng Xiaoping was formally brought back to the decision-making position for the third time in his career

foreign borrowing was still considered as reliance on foreign capitalists. Yet, the central government's decision to create SEZs made this possible and Guangdong made 'the project the linchpin of its strategy to develop both the special economic zones and the heartland.'[52]

Competing interests

The initiative and push for the Daya Bay project by the Guangdong government was balanced by various forces in the central government. Unlike the Qinshan project, the support for the Guangdong project in central government hinged on two issues: how to finance it and whose technology it would use. The country's economy was close to collapse. 'Mao's bias against foreign technology and foreign products had severely hurt China's modernisation, and per capita grain output in 1978 was the same as it was in the mid-1950s.'[53] China's near isolation meant the country had no foreign reserves to finance its necessary imports of advanced technology and equipment that would facilitate a rapid start to the reform. In 1978, China faced a growing trade deficit, despite increase in income from its petroleum exports. Clearly new and increasing sources of foreign exchange had to be discovered. If the government wanted to provide more than 40% of Chinese villages, with a population of about 300 million peasants with access to electricity, it required a large amount of capital, something unfortunately the country did not have.

In 1977, the SPC submitted a plan to the State Council on importing the whole-set equipment to speed up the four modernisations. It estimated that in the next eight years (up to 1985), China would need at least $6.5 billion in foreign exchange. In 1978, at a State Council meeting, line ministries reported that they would need to import at least 22 projects, which would be worth $7.8 billion. In 1978 alone, the country would need $1.2 trillion in foreign exchange. Yet, in 1977, the total exports were $7.95 billion and imports were $7.21 billion. China's trade in 1978 accounted for only 0.75 % of total world trade.[54] In 1978, China had a total foreign exchange reserves of only US$167 million and by 1980, it had gone down to a deficit of US$1.36 billion. A limited amount of total trade and especially foreign exchange reserves would not be able to support the necessary imports.

To finance a large scale of imports, the Bank of China had adopted aggressive policies to attract depositors at its overseas branches. It also borrowed at the European financial markets at an interest rate of 15–16 %. The estimated total cost of $5.1 billion for the Daya Bay project meant a heavy financial burden for the country, while the investment would not see an immediate return or even an immediate impact on alleviating power shortages.

Following the decision made at the 3rd Plenary Session of the 11th National Congress of the CCP to start economic reform and the opening up of the country, the Chinese minister of Foreign Trade announced at a press conference in Hong Kong on 15 December 1978 that the Chinese government had abandoned its traditional two restrictions: (a) no sovereign borrowing – the government would not borrow or guarantee any foreign borrowing; and (b) no foreign direct investment.[55] The two restrictions were based on the belief that the country should not have domestic and foreign debts. This was a significant change in ideological thinking and in practice: it was now possible to use the capital from capitalists to develop a socialist economy. When asked, 'What would be left of Communist ideology in China' if China depended on foreign capital and technology for its nuclear development, then Premier Zhao Ziyang replied:

> I have complete confidence in the ideas of Marx and Lenin. But I deal with the concrete situation in which China finds itself. China would seek to take what was good from the West, keep out the bad.[56]

In the following year, the State Council confirmed this significant change by issuing a series of directives that allowed selected provinces and cities to bypass the Ministry of Foreign Trade and deal directly with foreign firms, while retaining a share of foreign exchange earnings. Selected provinces and state-owned enterprises were also allowed to receive bank loans as part of their investment rather than strictly relying on budget allocation. It was a tentative and cautious step because even though foreign borrowing would be encouraged for key projects, it would have to be based on the principle of the ability to service its debt. If one could not demonstrate the ability to repay the debt, one would not be allowed to borrow.

Meanwhile, as a measure to open its economy, the central government also announced its policy on foreign investment. Foreign companies that wanted to move their processing industries to China, initially to special economic zones, could do so and China would offer cheap land, cheap labour and a stable environment for the investment. Finally, the Ministry of Finance submitted a report to the State Council that illustrated ways for short-term borrowing from overseas markets to address overall balance requirements.[57]

These policy changes made it possible for the Guangdong government to approach businesses in Hong Kong, in this case, Hong Kong's utility company, China Light & Power Co. Ltd. (hereafter called China Light), to form a joint venture in the autumn of 1979. This was the first joint venture in China. Despite the major reform measures adopted following the decision made by the Politburo in 1979, debates on financing the Daya Bay project remained controversial.

At first, many officials at the central government felt uncomfortable working with 'capitalists'. 'Maybe we should scale down the size of the project and then we could finance and build it on our own and then we could sell the electricity to Hong Kong to earn foreign exchange,' suggested the head of the nuclear department of the MEP. The idea of working with Hong Kong was at the heart of Deng's idea of 'opening up' and at the centre of the Guangdong government's plan. The advocates argued, with the participation of China Light, the joint venture would be able to access international finance, technology and managerial skills. It would also be able to sell no less than 50% of its electricity back to Hong Kong as a way to pay the debts. Moreover, to make this project economically viable, it had to have a large size, which was financially and technically impossible at the time in China. This would mean a fundamental shift from self-reliance to whole-plant imports. The plan drew immediate criticism from all directions, mainly from those who were concerned about the ability to finance the system and those who insisted that the project should be based on the principle of self-reliance, as instructed by Zhou Enlai.

The resistance from the SPC, the Ministry of Finance and some senior officials continued. Chen Yun (陈云) and many others did not oppose the project *per se* because they held the view that the project was achievable with a joint venture, foreign borrowing and through the future sale of electricity to Hong Kong. They were, nonetheless, concerned that China might have 'only a limited capacity to absorb foreign capital' and that cooperation should proceed carefully, keeping risks at a minimum.[58]

Regarding the ability to finance the project, several high-level meetings were held, attended by ministers from line ministries and various state commissions. The statement made by Yu Qiuli (余秋里), vice-premier of the State Council, in 1981 well reflected the mentality at the time:

> I support the nuclear energy development program because it was approved by Zhou Enlai and Deng Xiaoping, but given the current economic situation, I would say it is better for us to expand our thermal generation capacity than develop a nuclear power project; if we have to have a nuclear power plant, it is better for us to build one than purchase one from overseas.[59]

Others made their case. The prevailing view was: 'China has a lot of coal and hydro potential and we should allocate the resources to develop this potential rather than use precious foreign exchange to purchase foreign reactors'. It was not only different ministries and commissions that were fighting for their projects to be financed, the competition was also among different provinces. At the time when power shortages spread across the country, 'deciding the size, type, and location of China's new power plants was an extremely politicised subject' with localities fighting for as much electricity as they could get and coal power, hydropower and nuclear interests competing for limited funds.[60]

In early 1981, the SPC asked the MEP to submit a comparative study on the proposed NPP in Guangdong and a hydropower project in Guangxi along the Hongshui River. The MEP was in favour of the hydro project because it could be completed at a lower cost and more rapidly than a nuclear project to deal with immediate challenges of power shortages. The SPC and Ministry of Finance raised concerns about high costs of NPPs but did not openly oppose the project because Deng and the Politburo approved the project in principle.

Knowing the reservations of many in the central government, officials from Guangdong made frequent trips to Beijing to lobby the central government. In May–June 1981, a series of meetings were held in Guangzhou and Beijing to discuss the nuclear project. They were sometimes attended by more than 250 people from some 20 institutions, including those from the SPC, MWREP, Ministry of Finance, Bank of China, COSTIND and Tsinghua University. No consensus was achieved on the issue. At one of the meetings called by the State Council and chaired by the SPC, eleven ministers and commissioners discussed the feasibility report of the Guangdong NPP project. There were seven items on the agenda: the types of reactors to be built, the grid connection, safety, equipment, siting, laws and regulations, and economics and finance. A general consensus was achieved on six of the seven items, the exception being economics and finance. The feasibility report was rejected at this meeting because the SPC refused to give the green light on economic grounds. After the meeting, Gu Mu repeated the message to the participants that the issue was not whether China would start a nuclear energy programme or not, but how it would carry it out.

Others were worried about the message and implication to other provinces and regions that Guangdong was receiving 'special treatment' from the central government if the project went ahead. Given that the project might cost as much as US$4 billion or more and the country had only a limited amount of resources for large projects, 'should we give it all to Guangdong?' some asked. When Fujian, the other province that was allowed to set up special economic zones, asked for more power generation projects, Li Peng, vice minister of MWREP, told its governor that the province should expand its thermal generation capacity rather than think about nuclear projects. His colleagues, however, suggested that this might not be a wise suggestion given the provincial rivalry.

Another contentious issue was whether China should import a turnkey NPP station or it should rely on its own technology and human capital and import only some of the components. Guangdong pushed for a turnkey project because, its officials argued, China did not have the technology or human capital to build a nuclear power station. It would take more than two decades to develop these capacities and Guangdong could not, and did not want to, wait. Some argued against importing a turnkey project as a route for nuclear energy development in China because it would use up much of the country's hard-earned foreign exchanges and squeeze out its own industries; it would be better to invest the resources in developing its own technologies and human resources. Others argued that, even though it would be expensive, resources would be better used because China could introduce the most advanced technology for NPPs with the estimated 40–60 year life span. In addition, the argument went that China had already fallen behind the West in nuclear development, and it was critical to 'actively import advanced equipment, advanced technologies and advanced management experience, as well as funds and personnel.'[61] With imported advanced technology, the Chinese nuclear community could digest, absorb and master it and then build its own more advanced brand with further research. During the next three decades, this argument was to be repeated by those supporting technology imports.

Wang Ganchang, one of China's leading scientists and trained in Germany, along with some of his colleagues in the nuclear establishment, opined that China should be ready to import foreign technology and equipment for larger nuclear power plants. In November 1980, Wang told the American Nuclear Society that China would implement a nuclear energy programme and the only question was to what extent it would seek foreign technical assistance:

> We might seek technical help from a friendly country well-advanced in nuclear power and undertake with her an all-out cooperative program, like that between Brazil and the Federal Republic of Germany, so as to realize technical transfer in the shortest time, leading to a capacity of designing, building, and operating a commercial nuclear power plant by ourselves in the 1990s. As an alternative, we might also choose to rely mainly on self-reliance with a limited amount of technical help from abroad.[62]

The position taken by many scientists was attacked by those who would have liked to see the resources put into the development of domestic technology and equipment. Zhang Aiping (张爱萍), the minister of COSTIND, wrote to Chen Yun asking for his support for the position that China should develop its own nuclear energy programme rather than import a turnkey project: 'We support importing advanced technology, but we oppose relying on foreign technology for our nuclear energy development,' stated Chen Yun. Zhang represented those who had been involved in the nuclear weapons programme and believed that the sector needed the full support of the government.

The potential import of foreign technology was also seen as a threat to the domestic industries. On 19 June 1982, as Li Peng recalled, Zhang Jinfu (张劲夫), the State Council councillor in charge of finance and economy, passed him a letter from a foreman at a Beijing Boiler Factory, accusing the MWREP of being traitors by negotiating with France on importing a nuclear power station. Some people from the Shanghai Boiler Factory sent letters directly to the Politburo. Zhang Jinfu had been involved in the nuclear weapons programme in the 1950s and 1960s and had been beaten up during the Cultural Revolution for making similar accusations. He understood the implication of making these accusations and passed on the letter as a precaution.[63]

These two issues – financing and technology – continued to be at the centre of debates well into the mid-1980s. The opposition to borrowing on international markets was strong. A joint venture was considered to be selling out national interests, while purchasing whole plants was strongly resented by conservatives and by some officials at key factories building generators and turbines.

For the proponents, the project represented a new era for the country to open up to the outside world and take advantage of the capital and technology that could be supplied by the developed countries. One of the negotiators for the Daya Bay project described foreign borrowing as 'borrowing a chicken to lay eggs for China'. In May 1982, at a State Council meeting, heated debates broke out on the potential financial as well as physical risks of the project.

Finally, the participants reached several tentative agreements: (a) there were risks involved in all nuclear power projects no matter where they were, but the risks were manageable; (b) negotiation should take place with Hong Kong on nuclear power projects; (c) it was acceptable to borrow to build domestic infrastructure and the borrowing would require some level of imports; importing goods from foreign countries would not necessarily mean being subservient to these countries; (d) if Guangdong would not grab the opportunity, Hong Kong would go elsewhere to build power projects and (e) there were disadvantages as well as advantages in having a joint project – interest on borrowings would rise and so would electricity prices, because capitalists would not invest without profits. These basic understandings mirrored changing ideas at the time, as the country moved from a rigid, closed and autarchic economy to a more flexible, open one.

Based on these common understandings, the State Planning Commission agreed that: (a) the Guangdong provincial government could start negotiating with their foreign counterparts, and if required, it could bring MWREP into the negotiation. However, at this stage it was not a formal bilateral negotiation; (b) the State Council would form a negotiation coordination group with members from the SPC, MWREP, Bank of China, the Ministry of Finance, MNI and other relevant ministries, but the group would be led by MWREP, not the SPC; (c) the negotiation team would approach as many of its foreign counterparts as possible to create a sense of competition, but China would not open its bidding at that time and (d) a formal message would be sent out to the international community that China was serious about its nuclear energy programme. A vice commissioner from the SPC emphasised that even though the team could now start approaching its foreign counterparts, this was not going to be a formal bilateral negotiation because the SPC still had strong reservations about financing and technology imports.

In the following months, the debates continued. Various delegations were sent to visit nuclear power stations in Britain, France, Finland and some Eastern European countries. In December 1982, the State Council formally approved the construction of a nuclear power station in Guangdong province, but no decision was made on how it would be financed, whose technology it would use or who – MNI or MWREP – would lead the project.

Given that the first nuclear project was to be constructed primarily for selling electricity to Hong Kong for foreign exchange, the site had to be close to Hong Kong. It had to be on the coast and in a SEZ so that it could take advantage of the special policies. It was quickly decided that Daya Bay, about 45 km from Shenzheng, then a small village at the border to Hong Kong, and about 50 km from Hong Kong, was to be the site. Daya Bay was chosen to kill two birds with one stone – to supply the power hungry province with much needed electricity but, more importantly to bring Hong Kong in on the deal and thereby offer a more tangible gesture than Vice Chairman Deng Xiaoping's verbal assurance in April 1979 that investors in Hong Kong – worried about the territory's future – 'should put their hearts at ease.'[64] The State Council also decided upon the principles of the project: 'loans for construction, power sale for debt payment, and joint management' (借贷建设, 售电还贷, 合资经营).

International cooperation

China started its nuclear energy programme at a time when it had just opened up to the outside world and it was important to build good diplomatic relationships with other countries. The Daya Bay project was approved at a time when the nuclear industry in developed countries had suffered its first setback after the Three Mile Island accident and faced an increasingly organised anti-nuclear movement. International vendors were competing for business outside their countries just to survive. This meant that political and diplomatic considerations became part of the general background for the negotiation with nuclear vendors. Meanwhile, Chinese decision makers were being lobbied by international companies, whose governments were backing bids for business both in and outside China.

Chinese nuclear scientists were part of the first groups of scientists to visit the US and to have exchange programmes with their American colleagues. Chinese scientists preferred the commercially successful PWRs produced by Westinghouse, while American businesses were also keen to pursue business opportunities with China. To get their foot into the Chinese market, some American Chinese lobbied Li Peng and other leaders. One long-term American friend of Li Peng and some leaders in China told the Chinese government that the Tennessee Valley Authority (TVA) had two units of 130MW PWR in storage and, if China wanted them, TVA had the political support from the Republican leader in the Senate. Li Peng's friend suggested that even if in the end the US would not sell the units to China because of restrictions of the Nuclear Non-Proliferation Treaty (NPT), just getting the deal to the discussion stage would allow China to bargain with the French.[65] Knowing that there was little hope of getting nuclear technology or equipment from the US, the new deputy minister of Foreign Affairs in China was delegated to help build up economic contacts with other countries and to help negotiate a nuclear deal.

Long before the Chinese made a formal decision on a turnkey project in Daya Bay, scientists in China tried to convince their government that American technology was to be preferred. Deng wanted to build up a relationship with the US, and negotiations between China and the US over a nuclear cooperation agreement started in 1981. 'In January 1983, US officials negotiating a nuclear cooperation agreement with China linked possible US nuclear exports to China with its reported nuclear proliferation practices, particularly in Pakistan.'[66] Without changing its position on nuclear non-proliferation, China submitted its application to join the International Atomic Energy Agency (IAEA) in 1983. This was at a time when China was learning how to be a member of an international community. It joined the World Bank and the IMF in the same year and started seeking assistance from these institutions, including the IAEA.

The Chinese expressed their strong desire to acquire American technology and Westinghouse wanted to sell its reactors to China. In 1982, when 'discussing with the Chinese on the possibility of an agreement for peaceful nuclear cooperation, which would enable us to compete commercially in the development of China's nuclear power program', US State Department officials told the Chinese that they would have to provide assurance on the peaceful use of the nuclear technology. The US would also demand a guarantee against future re-export if nuclear fuel were to be provided. The Chinese told the Americans that 'they would use such technology only for power, but they had problems with allowing international inspection.'[67] This led to an agreement between the US and China that President Ronald Reagan brought back to Washington after he had visited Beijing in 1984. The agreement was rejected by the Congress.

Meanwhile, in Beijing, foreign affairs officials lobbied on behalf of the countries to which they were appointed. The Chinese ambassador to Moscow tried to convince those at MWREP and MNI that the Soviet Union had mature technology and China already had experience in building heavy water reactors. If China could make a deal with the Soviet Union, it would help improve bilateral relations and help China's position in Southeast Asia, especially over Vietnam and Cambodia. By and large, their lobbying was ignored because many in the central government did not think the relationship with the Soviet Union would improve in the short term.

The Chinese ambassador to France pushed for the deal with the French. When French Premier Raymond Barre visited China in January 1978, he took with him the two chairmen of Alsthom, a large French multinational conglomerate. In December 1978, as part of a seven-year trade agreement, Deng announced that China would import two nuclear reactors from France, valued at $14 billion.[68] The sale hinged on the Chinese agreement that the technology would only be used for peaceful purposes and be subject to the inspection of IAEA. This was at least the understanding of the Americans because when the French company Framatome used technology licensed by the Westinghouse Electric Corp. in constructing its atomic power plants it agreed any future sale would have to be approved by the White House. This was just before Deng's announcement that 'the White House issued a statement in late November [1978] saying it would not oppose the sale if France and China would agree to ensure that plutonium is not extracted from spent fuel to use in nuclear weapons.'[69]

Given the restrictions of the US Nuclear Non-Proliferation Act that forbade the sale of nuclear power equipment to countries that had not agreed to international inspection of their installations by the IAEA, the choice for China would be between the French Framatome and West Germany's Kraftwerk Union – a joint venture between Siemens and AEG. Both had substantially modified the Westinghouse PWRs, which they claimed they would be able to export without American permission.

Hong Kong's China Light wanted the British to win the deal and the Thatcher government believed it held an advantage with its control over Hong Kong. Yet after spending millions of pounds since World War II developing various kinds of nuclear reactors, Britain had not been able to build one that was regarded as being suitable for export. Nevertheless, the British government was determined to get involved in the project:

> Although it cannot hope to supply an export reactor at this stage, Britain would like to supply the associated turbine island consisting of turbine-generators and auxiliary equipment. These could be valuable orders running into hundreds of millions of pounds and would provide several years of work to the still order-short British process plant industry as well as numerous jobs to help alleviate the country's serious unemployment problem.[70]

Sir Lawrence (later Lord) Kadoorie, chairman of China Light, had every intention of keeping the British in the game. Right after the idea of forming a joint venture between Guangdong and China Light, Sir Lawrence requested a feasibility study of a nuclear power plant be done conjointly by the British Department of Industry, the Central Electricity Generating Board and the Atomic Energy Authority. The feasibility study was completed in late 1980, which made it possible for China Light to go ahead with the negotiation on the project with their Chinese counterparts. It also ensured a place for British General Electric Corp (GEC).

The CCP Central Committee and the State Council decided that nego-tiations on the nuclear project and negotiations with the British govern-ment on the return of Hong Kong after 1997 were two separate matters: one was between the Guangdong power sector and a Hong Kong corpora-tion; the other was between two sovereign states. The two issues nonethe-less became entangled throughout the prolonged negotiations. For example, a deputy minister of Foreign Affairs in China warned her colleagues from the MWREP not to push for the deal because the gap between China and Britain over the Hong Kong issue was huge and the result of these negotia-tions might affect the Daya Bay deal. China wanted a return of sovereignty, to retain the capitalist system and special management of Hong Kong, while Prime Minister Margaret Thatcher wanted to keep the status quo with an extension of the lease.

By the end of 1982, it became a three-way negotiation between the Chinese (joint efforts between the central government represented by the MWREP and the Guangdong government), the Anglo-French joint force, and China Light and Britain. The Chinese wanted to purchase the reactors, which were eventually provided by France, and turbine generators, which were provided by the British, they also insisted on technology transfers and loans with preferential conditions and low interest rates as conditions for the agreements.

In May 1983, French president Francois Mitterrand visited China and secured from the Chinese another memorandum of understanding 'ensur-ing a substantial participation by France in the Chinese project.' British officials reacted immediately, claiming: 'We are not aware of any serious approval of this nature and we still believe the Chinese are seriously con-sidering signing contracts with GEC.'[71] Indeed, in March 1983, the British delegate led by the Deputy Secretary of Industry signed a similar memo-randum regarding the supply of power-generating turbines by GEC.[72] The final agreements were signed in 1984 and the Daya Bay Project started its construction in 1987, ending a decade of learning experience in dealing with the outside world.

Conclusion

From the early 1950s to the 1970s, nuclear programmes had been the priority in terms of political support and allocation of finance, physical and human resources. It was under the tight control and subject to the strict instruction of the CCP Central Military Commission and the Politburo. Nuclear weap-ons programmes could not escape the political turmoil that had swept the country for two decades. Many scientists were protected by those in charge of the weapons programmes, especially by the military generals who moved them to the Gobi Desert or other remote areas close to testing places, but also away from the epicentre of politics in Beijing.

The basic research on reactors (civil and submarine alike) was not so fortunate. It competed with the weapons programme for resources and political attention and was kicked like a football from one organisation to another, none of which could spare resources for it.

The internal debate and competition for priority in the nuclear sector continued after the 1970s when China decided to pursue a nuclear energy programme. It took about a decade for China to obtain its first commercial nuclear power station. The construction of Daya Bay started in August 1987. This indicated the beginning of the Chinese nuclear programme, and also the beginning of the reform in the energy sector. Daya Bay was the first joint venture in the energy sector after the central government removed its restrictions on non-foreign borrowing and foreign investment.

How the Qinshan and Daya Bay projects were initiated and developed also reflected politics in China – both were pushed by special or local interests rather than by a set of coherent policies developed by the central government. Guangdong not only obtained a nuclear power plant but also the autonomy it had demanded to start a new path to economic development – market rather than planned, open rather than closed, and integrating with international economies rather than existing in isolation from them. The two projects – one an example of self-reliance and the other an imported turnkey project – set the two distinct paths of China's nuclear development in the following three decades.

Notes

1. Zhang Youxin, 'Daya Bay Nuclear Power Plant,' *People's Discussion Forum* No. 26, May 1994, pp.46–48.
2. Pollack, 1972.
3. 孟戈非, 2002, pp.176–78.
4. Johnston, 1995/96, p.8.
5. Lewis and Xue, 1988.
6. Ibid, p.54.
7. Li, 2007, p.168.
8. Lewis and Xue, 1988, p.64.
9. 当代中国的核工业, 1987.
10. Naughton, 1988.
11. 孟戈非, 2002; Shambaugh, 1987.
12. Gourievidis, 1985.
13. For the debate and power struggle, see 孟戈非, 2002.
14. Jiang Shengjie, 'Developing China's Nuclear Power Industry', *Beijing Review*, 27:25, 18 June 1984, pp.17–20.
15. Naughton, 1993; Robinson, 1982.
16. 刘国光, 2006; Chan, 2003; Gittings, 2005.
17. Fang I, 'China's New Priorities for Technology Development,' *China Business Review*, 5:3, May–June 1978, p.5.

18. Ibid, p.5, p.41.
19. Anonymous, 'Exporter's Notes,' *China Business Review*, May–June 1978, p.41.
20. Quoted from Fountain, 1978, p.39.
21. Anonymous, 'Proposal for Nuclear Power Stations,' *Beijing Review*, 23:4, 8 December 1980, p.4.
22. Quoted from Vaclav Smil, 'Deep Structural Deficiencies', *The China Business Review*, January–February 1980, p.65.
23. Quoted from Fountain, 1978, p.38.
24. Ibid.
25. Anthony Rowley and Pauline Loong, 'The Politics of Nuclear Power,' *Far Eastern Economic Review*, 10 October 1980, p.48.
26. 李鹏, 2004, pp.25–26.
27. Jiang 1984, p.139.
28. Gao Bo, 'Improve the Country's Energy Situation', *People's Daily*, 12 November 1979; Yang Zhirong, Zhu Bin and Zhang Zhengmin, 'Several Technology Issues in Energy Development', *People's Daily*, 28 February 1980; Chang Jiming, Zhao Zhenggi and Chen Fumin, 'Nuclear Power Plants: a new force for the electric power industry', *Zhejiang Daily*, 31 August 1982.
29. Jones, 1981, p.32.
30. Conroy, 1989.
31. Jones, 1981, p.35.
32. Jiang 1984.
33. Reardon-Anderson, 1987, p.36.
34. Lieberthal and Oksenberg, 1988, p.94.
35. Naughton, 1993, p.500.
36. Lieberthal and Oksenberg, 1988, p.35.
37. Ibid., p.406.
38. Reardon-Anderson, 1987, p.40.
39. Anonymous, 'Proposal for Nuclear Power Stations', *Beijing Review*, 8 December 1980, p.3.
40. Jones, 1981, p.32.
41. Jiang 1984.
42. Reardon-Anderson, 1987, p.41.
43. Quoted from Jones, 1981, p.35.
44. Lieberthal and Oksenberg, 1988, pp.22–23.
45. Weil, 1982, p.41.
46. Ibid, p.41.
47. Gourievidis, 1985, p.110.
48. Zheng, 2007, p.239.
49. Vogel, 1989.
50. Quoted in Naughton, 1993, p.509.
51. 'Interview of Gu Mu,' *Nanfan Daily*, August 9, 2004.
52. Anonymous, 'A Nuclear Hong Kong?' *Asiaweek*, July 20, 1986, p.31.
53. Perkins, 1994, p.23.
54. Naughton and Lardy, 1996.
55. 刘国光, 2006, p.436.
56. E.J. Dionne, Fr., 'China and France Near Nuclear Accord', *New York Times*, 3 June 1984.
57. Nicholas Lardy, *Foreign Trade and Economic Performance in China, 1978–1990*, Cambridge: Cambridge University Press, 1992.
58. Lieberthal and Oksenberg, 1988, pp.226–27.

59. 李鹏, 2004, p.34.
60. David Denny, 'Electric Power and the Chinese Economy', *China Business Review*, 12:4, July–August 1985, p. 18.
61. Peng Shilu, 'The Role of Nuclear Power in China's Power Structure,' *Nuclear Power Engineering*, 6:6, 1985, pp.1–3.
62. Quoted from Dori Jones, 'Nuclear Power: Back on the Agenda', *China Business Review*, 7, 1981, p.35.
63. 李鹏, 2004, p.45.
64. Anthony Rowley and Pauline Loong, 'The Politics of Nuclear Power,' *Far Eastern Economic Review*, 10 October 1980, p.48.
65. 李鹏, 2004, p.110.
66. Kan and Holt, 2007, p.4.
67. Bernard Gwertzman, 'US and China Discuss Export of Nuclear Technology,' *The New York Times*, 2 June 1982.
68. 'Power,' *China Business Review*, November–December 1978, p.51.
69. Ibid.
70. Anthony Rowley and Pauline Loong, 'The Politics of Nuclear Power', *Far Eastern Economic Review*, 10 October 1980, p.48.
71. Mark Baker, 'French in China-N Station Deal', *Financial Times*, 6 May 1983, p.20.
72. Robert Delfs, 'The Balance of Power', *Far Eastern Economic Review*, 19 May 1983, p.80.

3
Expanding the Nuclear Energy Programme

After the two nuclear power projects were initiated in the late 1970s and early 1980s, nuclear energy development in China stalled, primarily because there was no coherent policy from the central government, no voice speaking on nuclear energy in high places, and no institution at the central level willing and able to host its development. Nuclear energy became a subject driven by political and diplomatic concerns and specific and local interests. It did not become an energy issue until the 2000s, when power shortages hit two-thirds of the provinces from 2002 on and environmental pollution, because of a heavy reliance on coal, became a more pressing challenge for the Chinese government. Looking for alternatives led to a new call for expanding nuclear energy in China. In March 2005, Premier Wen Jiabao said at a meeting of the Standing Committee of the State Council that 'China needs to change its structure of its electricity generation; expand its hydro capacity, optimise its thermal development, actively promote nuclear energy, appropriately develop gas-fired electricity and encourage renewable energy' (调整电源结构，大力开发水电，优化发展煤电，积极推进核电，适度发展天然气，鼓励新能源发电). By June 2010, China had 11 nuclear reactors in operation and another 24 under construction (see Table 3.1).

The new enthusiasm for nuclear energy expansion in China placed new challenges on resources allocation, technology selection, waste management and safety regulation. By and large, however, the development was pushed by provinces and the major players in the nuclear and power sector. In 2008, when the National Energy Administration was created, the nuclear sector was placed for the first time under the umbrella of energy policy-making.

Stalled development in the 1990s

In the 1990s, four NPP projects received approval and went into construction: Qinshan III in 1994, Qinshan II in 1995, Lingao in 1995 and Tianwan in 1997. Each of these four projects had been pushed by different players

Table 3.1 Nuclear power reactors in China, as of June 2010

Name	Type	Status	Capacity (MWe) Gross	Location	Owner	Construction starting date	Date connected
DAYA BAY-1	PWR	Operational	984	Guangdong	GNPC	1987/08/07	1993/08/31
DAYA BAY-2	PWR	Operational	984	Guangdong	GNPC	1988/04/07	1994/02/07
LINGAO 1	PWR	Operational	990	Guangdong	GNPC	1997/05/15	2002/02/26
LINGAO 2	PWR	Operational	990	Guangdong	GNPC	1997/11/28	2002/12/15
QINSHAN 1	PWR	Operational	310	Zhejiang	CNNC	1985/03/20	1991/12/15
QINSHAN II-1	PWR	Operational	650	Zhejiang	CNNC	1996/06/02	2002/02/06
QINSHAN II-2	PWR	Operational	650	Zhejiang	CNNC	1997/04/01	2004/03/11
QINSHAN III-1	PHWR	Operational	700	Zhejiang	CNNC	1998/06/08	2002/11/19
QINSHAN III-2	PHWR	Operational	700	Zhejiang	CNNC	1998/09/25	2003/06/12
TIANWAN 1	PWR	Operational	1000	Jiangsu	CNNC	1999/10/20	2006/05/12
TIANWAN 2	PWR	Operational	1000	Jiangsu	CNNC	2000/10/20	2007/05/14
CHANGJIANG 1	PWR	Under construction	1087	Hainan	CNNC	2010/04/25	
FANGJIASHAN 1	PWR	Under construction	1087	Zhejiang	CNNC	2008/12/26	
FANGJIASHAN 2	PWR	Under construction	1087	Zhejiang	CNNC	2009/07/17	
FUQING 1	PWR	Under construction	1087	Fujian	CNNC	2008/11/21	
FUQING 2	PWR	Under construction	1087	Fujian	CNNC	2009/06/17	
HAIYANG 1	PWR	Under construction	1250	Shandong	CPI/Shandong	2009/09/24	
HAIYANG 2	PWR	Under construction	1250	Shandong	CPI/Shandong	2010/06/21	
HONGYANHE 1	PWR	Under construction	1080	Liaoning	CPI/CGNPC	2007/08/18	
HONGYANHE 2	PWR	Under construction	1080	Liaoning	CPI/CGNPC	2008/03/28	

continued

Table 3.1 continued

Name	Type	Status	Capacity (MWe) Gross	Location	Owner	Construction starting date	Date connected
HONGYANHE 3	PWR	Under construction	1080	Liaoning	CPI/CGNPC	2009/03/07	
HONGYANHE 4	PWR	Under construction	1080	Liaoning	CPI/CGNPC	2009/08/15	
LINGAO 3	PWR	Under construction	1087	Guangdong	CGNPC	2005/12/15	2010/08/31
LINGAO 3	PWR	Under construction	1086	Guangdong	CGNPC	2006/06/15	
NINGDE 1	PWR	Under construction	1087	Fujian	GNPC	2008/02/18	
NINGDE 2	PWR	Under construction	1080	Fujian	GNPC	2008/11/12	
NINGDE 3	PWR	Under construction	1080	Fujian	GNPC	2010/01/08	
QINSHAN II-3	PWR	Under construction	650	Zhejiang	CNNC	2006/03/28	
QINSHAN II-4	PWR	Under construction	650	Zhejiang	CNNC	2007/01/28	
SANMEN 1	PWR	Under construction	1115	Zhejiang	CNNC	2009/04/19	
SANMEN 2	PWR	Under construction	1115	Zhejiang	CNNC	2009/12/17	
TAISHAN 1	PWR	Under construction	1750	Guangdong	GNPC/EDF	2009/10/28	
TAISHAN 2	PWR	Under construction	1750	Guangdong	GNPC	2010/04/15	
YANGJIANG 1	PWR	Under construction	1087	Guangdong	GNPC	2008/12/16	
YANGJIANG 2	PWR	Under construction	1087	Guangdong	GNPC	2009/06/04	

Source: IAEA, 'Power Reactor Information System', available at http://www.iaea.org/programmes/a2/.

and approved for different reasons. In general, however, their approvals mirrored the political and economic development in the country at the time. Qinshan II was approved because, as the Chinese premier Li Peng said repeatedly in 1995, 'the whole nuclear industry would collapse if we do not give it some projects.'[1] Lingao was pushed by the China Guangdong Nuclear Power Corporation (CGNPC) that had owned and operated the Daya Bay NPP using its own profits to finance it. The Tianwan project was the product of bilateral diplomacy between Russia and China. Qianshan III, built in cooperation with Canada, was one piece of a whole bargaining process with the OECD countries, especially the US, to break the isolation that had followed 1989.

All four projects were made possible because the then premier Li Peng was a strong advocate and supporter of nuclear energy. This all changed in late 1990s when Zhu Rongji replaced Li and became the premier. Zhu was 'by leaps and bounds the more hard-charging reformer compared with Li.'[2] He had consistently stressed the need to break up government monopolies and to create competition in previously protected sectors. When he became the premier, Zhu quickly wielded a heavy axe on state-owned enterprises (SOEs) and presided over a 40% reduction in the public enterprise workforce. He did not believe it was economically beneficial for China to build nuclear power stations and was determined that no new NPPs would receive approval unless the nuclear sector, the China Nuclear Power Corporation (CNNC), reversed its loss-making situation. The Asian financial crisis offered him the opportunity and Zhu at the end of the 1990s ordered to freeze construction of all electricity generation projects, including nuclear power stations, for at least three years.

One paradoxical challenge facing nuclear energy development around the world is that the nature of nuclear energy development – intensive capital investment and long-life expansion – requires long-term planning and long-term stability in policies and commitment. However, nuclear energy development often faces two big uncertainties – turbulent politics and economic uncertainties. China faced both in the 1990s. In the end, political decisions prevailed. Because of the political nature of the decisions on all four NPPs, nuclear development in China fell into a pattern that many in the sector had warned against – different models of reactors to be adopted would make it difficult to bring the costs down and to regulate the industry. By the end of the decade, China had been building NPPs with technology from Canada, France, Russia and its own.

The normal impression of Qinshan I, II and III is that they must have been sequential projects. This, however, is not the case. Indeed, Qinshan III was approved before Qinshan II. They were three 'relatively independent' projects, owned by different institutional arrangements, operated by different institutions, with different models of reactors, and based on different technology.

The reactor adopted in the Qinshan I project was extrapolated from a Chinese nuclear submarine reactor design; Qinshan II used the Daya Bay PWR as a reference design and Qinshan III had Canadian CANDU heavy water reactors. They share the same name only because they are located in one large geographical area: Qinshan, Zhejiang province. The development of the three projects is a reflection of historical development and an indication of a lack of central planning and long-term strategic thinking when projects were put on the agenda.

Qinshan III: Canada showed its intention to sell CANDU reactors to China as early as the 1980s, but China preferred the American PWR to the HWR technology on which CANDU was based. When the Daya Bay project was approved and went under construction, China indicated that it would build more NPPs and it became clear that the US would not and could not sell nuclear technology to China. Canada approached China again, and in 1988, the governments in Ontario and Jiangsu province signed memoranda of understanding (MOU) on energy cooperation. One of the items on the agenda was to assess the potential for Canada's CANDU reactor. 'Government, power, and nuclear industry spokesmen from both sides of the Pacific say Canadian-Chinese nuclear trade is both feasible and likely in the near future.'[3] Some Chinese were interested in this cooperation because they were familiar with heavy water reactors and liked the fact that with them they would be able to burn recycled PWR fuel, while the Canadians were eager 'to market Canada's nuclear technology abroad.'[4] In 1989, just when the two sides were close to a breakthrough in their negotiations for potential cooperation, the negotiations collapsed when Canada along with others imposed an embargo against China.

After almost a decade of on–off negotiations, serious negotiation was resumed only after Deng's 'Southern Tour' that brought the reform back to life. In November 1994, the Canadian Prime Minister, Jean Chretien, and Chinese Premier Li Peng signed a Nuclear Cooperation Agreement to assure compliance with IAEA non-proliferation terms. The negotiation was also shifted from the MEP and the MNI to the Ministry of Energy and CNNC. In the following 20 months, hard bargaining took place mainly over project financing. Negotiations were also conducted to show the US and some OECD countries that Canada was more than willing to do serious business with China even if the US refused to do so.

The initial price tag set by Canada was $4.2 billion, while the Chinese government budgeted for only $2.4 billion. In addition to the large gap in the estimated costs, China also expected 'all financing (for the plant) should be from Canada at an interest rate lower than that set by the OECD', which was 8.85% at the time.[5] During the negotiation, it was reported that the Canadian government had 'committed $1.5 billion in financing for the Qinshan project and AECL [Atomic Energy of Canada Limited] obtained commitments for additional financing from external partners, but the

Chinese still required more to make this viable.'[6] One explanation of the hard bargaining conducted by the Chinese was that the country had limited foreign reserves. In 1992, for example, the country had foreign exchange reserves of US$21.2 billion, but in the following year it ran trade deficits of US$12.2 billion. To finance such capital-intensive projects, the country depended on the financing that foreign suppliers could bring to the country. Meanwhile, with severe power shortages across China, few would want to wait for seven to ten years for electricity to be generated by a NPP.

Another explanation for the hard bargaining was that even though some in the nuclear sector preferred CANDU reactors because they were familiar with the technology and would have liked to continue using it, many more in the electricity industry and other government agencies preferred to expand PWRs because: (a) more than 60% of the world's nuclear generation capacity was based on PWR technology and there was a record of success; and (b) China had already imported Framatome's PWRs, which had been built based on Westinghouse PWR technology – an expansion of the same technology could bring down the costs and improve the safety record.

This debate among different players in China affected those sitting at the negotiation table, who were half-hearted some of the time while serious at other times. Finally, the international nuclear market at the time was a buyers' market and OECD countries had either stopped their nuclear power projects completely or simply banned further NPP expansion. Major international nuclear suppliers, such as CANDU, Framatome and Westinghouse, were competing in countries such as China.

When the project of Qinshan III was approved by the SPC on 26 February 1996 and the contract between CNNC and AECL was signed in November 1996, some commented that 'this deal with Canada would be a political answer to the US that says "to hell with your embargo".'[7] For many, while the project might not make a great deal of economic or technical sense, it did make a political statement to the US, which had imposed technology exports to China but from which China would prefer to get its technology. The contract became effective on 12 February 1997. CNNC created a subsidiary, the Third Qinshan Nuclear Power Company (TQNPC), holding 51% of its stake and the rest was shared among China Power Investment Corporation (20%), Zhejiang Power Corporation (10%) and another two investment companies of the Jiangsu (9%) and Shanghai governments (10%). TQNPC was the designated owner of the project while AECL was the main contractor.

AECL would build two CANDU 6 reactors with the capacity of 728MW, the only two heavy water reactors China has to this day. As the main contractor, AECL then subcontracted out to American Bechtel and Japanese Hitachi. All three were supported by the credits provided by the export–import banks in their relevant countries. AECL received $1.5 billion credits from the Export Development Corporation of Canada (EDC). The specific financial structure was: 71% of the total costs were provided by the three export credit agencies (made up of EDC 70%, US EXIM 16% and JEXIM 14%). These credits were provided following the OECD guideline of a 7.5% interest rate for 15 years. The justification for AECL to obtain assistance from EDC was that the deal would provide 15 000 jobs in Canada and the same argument was made by Bechtel and Hitachi. For the remaining 29% of the total cost, the State Development Bank guaranteed international borrowing (22%) and the remaining 7% was self-financing.

Among other things, TQNPC was responsible for preparing the site, providing permanent site facilities and local staff to the AECL Site Project Management Organisation, managing construction by sub-contract to Shanghai Nuclear Engineering Research and Design Institute (the former 728 Institute), managing licensing, providing the first fuel loan and initial heavy water fill. AECL, meanwhile, was responsible for providing the design and reactors, managing the reactor construction, and providing guidance and direction to TQNPC for commission. Bechtel/Hitachi was subcontracted to provide turbine generators. The construction of Qinshan III started on 10 March 1997 and its commercial operation began on 19 November 2002. The second unit was connected to the grid a year later. As the AECL predicted, the construction schedule was shorter than for PWRs.

Qinshan II: at the inauguration ceremony for the Daya Bay project, Li Peng declared, 'The conditions are fulfilled for the construction of the second phase of the [Chinese PWR] project.'[8] The idea of duplicating the PWR technology introduced from France for Daya Bay to develop its own more advanced reactors originated in the late 1980s. In 1988, CNNC and Jiangsu and Zhejiang provinces pooled their resources to plan for another NPP in either province and created what was later known as the CNNC Nuclear Power Qinshan Joint Venture Company Limited. The Beijing Institute of Nuclear Engineering and the Nuclear Power Institute of China in Chengdu, Sichuan province, were asked to design a 600MW PWR based on the model imported from France for the Daya Bay project. This was the origin of Qinshan II.

The project ran into trouble as soon as it was proposed. Following the Tiananmen incident in 1989, OECD countries imposed embargos against China. Without access to finance and foreign technology, the project had no hope of proceeding. The government also shifted its emphasis. In both 1990 and 1991, the State Planning Commission headed by vice-premier Zou Jiahua made it clear that capital investment would not go to new projects but would only go to those projects that could help with economic recovery. Qinshan II was put on the backburner. Even after Deng's 'Southern Tour', which revived the reform, the priority of the government was to prevent the economy from overheating. Capital investment was tightly controlled. New projects with the central government's funding were put on hold. Meanwhile, CNNC and its subsidiary put together a feasibility study, which was sent to the State

Planning Commission for approval in 1992. For the next four or five years, debate continued over several issues: (a) whether China had the resources to expand its nuclear energy programme at a time when demand for finance came from all directions and while the sources of finance were limited; and (b) whether Qinshan II should be based on indigenous technology or whether China should concentrate its efforts on one imported reactor-line, as France did in the 1970s, in order to speed up progress.

Some vice-premiers of the State Council supported the nuclear energy development in principle but emphasised that the country should tighten capital investments in order to control inflation; this should apply to Qinshan II too. In the long term, said Zou Jiahua, nuclear energy development was necessary because sooner or later fossil fuels would run out, but given the current macroeconomic situation it was too difficult to finance such a large capital-intensive project. However, Li Peng, the premier at the time, believed that Qinshan II should be given the green light in order to save the nuclear industry. Li Peng said, 'At the moment, it is very difficult for the nuclear sector because there are few government procurements; we approve this project under special conditions to save the whole industry.'[9] The SPC finally approved the project on 15 December 1995.

Both Daya Bay and Qinshan III were turnkey projects, while the Qinshan II project was billed as 'a step towards self-reliance'. The Chinese were supposed to build two 600MW PWRs with the Framatome's 900-MW PWRs installed in Daya Bay as a reference design. Because of differences in site conditions, capacity, grid connection and the desire to improve on the French design, engineers at the Beijing Institute of Nuclear Research had to make substantial and substantive modifications to the PWR model at Daya Bay.

The Chinese retained technical cooperation and assistance from France's Framatome and Electricité de France (EDF). Siemens AG's Kraftwerk Union (KWU) had wanted to get into the Chinese market since the early 1980s, because 'the days when KWU officials symbolically licked their chops over prospective nuclear power plant contracts in a host of countries' in Asia, Europe and North Africa had added.[10] They had placed their hope on China and the Soviet Union when the Chernobyl disaster destroyed the possibility of making deals within the Soviet republics. When Qinshan II was placed on the agenda, KWU was hoping to help design and build these two 600MW PWRs. The 1989 embargo brought this to an end.

When the project was back on track, CNNC signed a contract with Westinghouse in early 1995 to deliver two 650MW steam turbines to be used for Qinshan II. The US nuclear industry had long been lobbying intensely to free nuclear trade with China, while the experts in China had long expressed their desire to obtain the American technology. Indeed, the Chinese signed the agreement with Westinghouse when US Secretary of Energy, Hazel O'Leary, visited China. Before the Clinton administration could get congressional approval for the certification, however, Westinghouse could only sell the equipment to China that would be manufactured outside the US by Westinghouse licensees. Westinghouse itself, and its Spanish licensee Equipos Nucleares S.A., would supply steam generators and Mitsubishi Heavy Industries in Japan and would make the reactor coolant pumps and motors. The Chinese were supposed to assemble them. Because of this, CNNC also reached an agreement with Westinghouse for the engineering, project management and quality assistance support.

The Tianwan project was pursued for quite different reasons, mainly political and diplomatic rather than for economic, energy or technical. One of the items on the agenda when Soviet leader Mikhail Gorbachev visited China in May 1989 was potential cooperation on nuclear energy development, which was confirmed a year later when Li Peng visited Moscow. The idea was quickly pursued because of the political developments on both sides.

China became isolated once again after June 1989 when Western countries imposed sanctions and negotiations on several potential NPP projects stalled. Building a strong partnership with the Soviet Union was politically and diplomatically important for the Chinese government. Shortly after Gorbachev's visit, the Soviet Union started to crumble and the economy in most former Soviet republics collapsed, including that of Russia. The potential to build a NPP would have created an opportunity for Russia to keep one of its industries alive, while allowing China to start rebuilding its relationship with Russia now that OECD countries had shunned China. German KWU officials commented that the nuclear cooperation between the Soviets and Chinese was 'a logical consequence of the warming trend in Sino-Soviet relations.'[11]

In December 1992, China and Russia signed an agreement on cooperation in building NPPs in China. The agreement had been pushed through by China as a counter-response to the Bush administration's decision to approve leasing warships to Taiwan in July 1992. The supply of warships through a lease arrangement was seen as a violation of the principles set out in the Sino-US Communiqué, agreed upon on 17 August 1982. It was expected that with a newly elected president in office later in the year, Washington would 'step up pressure against foreign states seeking to intensify commercial relations with China' and Moscow was strongly urged by the Bush administration 'to exercise restraint in nuclear commerce with Beijing.'[12] To counter the US policies, China happily engaged in serious negotiation with its Russian counterparts on the nuclear energy project.

The project, however, received its share of criticism in China. Some liked the deal because a Soviet plant would 'be considerably cheaper than its Western equivalent, and a deal could be arranged on barter terms.'[13] Many others, however, especially nuclear experts, raised serious concerns about the safety standards of VVERs – the Russian version of light water pressurised reactors, especially after the Chernobyl disaster in 1986 when one of the four Russian-built VVERs was completely destroyed by fire. Finally, the political agenda overrode other considerations.

While people involved in the negotiations were asked to push through the deal as a 'political and diplomatic task',[14] safety concerns had to be taken into serious consideration. China demanded serious modification of VVERs and ordered that the new design be approved by the IAEA. Meanwhile, it was negotiating with Siemens to provide instrumentation and control equipment. Finally, it insisted that Chinese organisations would take charge of construction. Russia agreed on the last point but made sure that it would not pay for Chinese labour.

In addition to concerns about safety standards of those in the nuclear sector, or concerns about economic costs from those in the electricity sector and the province in China, both China and Russia were going through significant political changes. Government reorganisation took place in both countries in 1992 and 1993. On the Russian side, because the Ministry of Atomic Energy brought the military and electricity components together in 1992, officials from Minatom, the Russian nuclear power ministry, said that the ministry would 'not be directly involved in pending commercial talks' on the nuclear power project in China. The Russian firm Zarubeshatomenergostroy would be the party at the negotiating table.

On the Chinese side, the Ministry of Energy was dissolved and in its place several line ministries were created. Ministry of Electric Power was restored but not the ministry of nuclear industry. 'The Russian side is a ministry; which ministry should we have?' asked Li Peng, the then premier.[15] The Chinese government decided that the Ministry of Electric Power would take the lead of the Chinese team, which consisted of representatives from the power sector, CNNC and Liaoning province. In the middle of negotiations, in June 1995, the State Council decided to change the leader of the Chinese team from the Ministry of Electric Power to CNNC. 'The complex relationship among Chinese organisations involved' – CNNC, the Ministry of Electric Power and the Liaoning provincial government – undoubtedly was one of the factors for the prolonged negotiation, commented Western observers.[16]

The changing organisational arrangement highlighted a politically sensitive issue that had been lingering since the time the nuclear energy programme was placed on the agenda in China – who was in control? When the Ministry of Electric Power was the lead agency, it insisted that it would approve the site, the feasibility study and the long-term planning for the electricity connection to the grid before all these matters were submitted to the State Council for approval and before the next step could be taken in the negotiation. In sum, at every step, the Ministry wanted a say on all matters. This might be understandable given that electricity eventually had to be provided through a grid and that needed long-term planning. It created organisational jealousy because CNNC was 'a corporation' rather than a Ministry.

It was made known to the world that the site for this project would be near Dalian, Liaoning province. The Russians had done substantial work to modify the reactor designs to the Liaoning site without the protection of a formal contract. Two weeks after officials from CNNC told representatives from the European Commission about this project and the site; the State Council in September 1996 changed the site from Liaoning to Jiangsu. Speculation was made about the last minute change in the site's location. Some suspected that the change was a way for the Chinese to get out of the project contract because of their domestic concerns about the safety of Russian reactors. Some wondered whether this was an excuse for the Chinese to pull out because of the financial concerns. Even though the Russians offered a lower price for project than other Western vendors did, the Chinese had to come up with all the financing.

Some argued that it made sense for the Chinese to pull out of the deal because Li Peng had just announced that there would be no new nuclear power project for the 9th Five-Year plan (1996–2000). Some insisted that pulling out the deal was necessary because, with the background of Chernobyl, 'unless the reactors are built on a site already approved by the Chinese government, such as Qinshan, the local and regional authorities will have to be involved, and these won't likely favour construction of a Russian reactor on their territory.'[17] Others believed that changing the site was a rational decision made by the central government to meet rising electricity demand in places around Shanghai, while the northeast part of China, with its heavy industry, was under pressure to restructure. NPPs, argued many Chinese leaders, should be built at places where the energy endowment was low while economic growth was high so that it would be easier for the utility companies to up-adjust prices for electricity. Even though the GDP per capita in both Liaoning and Jiangsu was above the national average in the early 1990s, that of Liaoning was about 7% lower than that for Jiangsu. This might explain the reluctance for Liaoning province to help finance the project. It would make economic sense for relocating the site to Jiangsu.

With the Chernobyl disaster at its back, the Russians were in a disadvantageous negotiating position. After several changes, the Chinese pushed for further concessions. Given that both sides suffered shortages of foreign reserves in the early 1990s, the Tianwan project provided a good opportunity for them to make a barter deal work. Russia would supply 'all equipment and material' for the reactors and supply all the enriched uranium fuel of the VVERs, while China would pay for them with meat, eggs, clothing and the consumer goods that were in short supply in Russia, even though officials from Minatom said that they would expect China to pay for the two reactors in large part with foreign exchange. It was never made clear what specific financial arrangements had been reached between the two sides.

Once the State Council announced the site, all the agreements and contracts, the feasibility study, technology cooperation and the contract for the project itself were pushed through in record time.

In April 1998, the construction on the project was started. More problems emerged. For example, normally, a construction should not start until 60% of the blueprints are in place, but in the case of the Tianwan project, only about 6–7% of blueprints were in place when the project began. Delayed delivery of materials, equipment and blueprints was caused by a combination of factors, among which was the turbulent political situation in Russia where voucher privatisation in the early 1990s and then loans for share privatisation in 1995 to get Yeltsin re-elected completely threw the economy off balance. The producers could not meet the deadlines to deliver the products and when they did, the quality was always in question.

Both sides suffered from a shortage of experts. In Russia's case, many experts had left the field after the collapse of the Soviet Union, when they could not be paid. The reactor model used in Tianwan was AES-91, a modified version after the Chernobyl disaster. The Russians often did not know how to deal with the problems. Jiangsu Nuclear Power Corporation borrowed experts from other NPP projects around China for help. Meanwhile, with several NPPs under construction, there was a shortage of skilled labour everywhere.

In 1995, concerns about the safety of VVERs prompted CNNC to join the World Association of Nuclear Operators (WANO), an organisation formed after the Chernobyl disaster to improve safety by exchanging information and providing technical assistance. Daya Bay had already been a part of WANO because it is a two-PWR Framatome plant with management and operational assistance from EDF. CNNC joined partly because it would not be isolated by the nuclear community and partly because of the safety concerns in other NPPs. The triggering point was the Tianwan project.

Lingao: The development of the Lingao NPP project illustrates another aspect of the politics in the 1990s in China – the central and provincial relationship. While the Daya Bay project was under construction, Guangdong province and the joint venture created for the Daya Bay project were planning to buy another two 900MW-class PWRs, with technology transfer provisions from foreign vendors. The site was already chosen to be at Lingao, about 5 km from Daya Bay.

Framatome had long believed that it had positioned itself well with the arrangement of Daya Bay. When 'the award intention agreement' was signed in Beijing on 15 January 1995, the CEO of Framatome, Jean-Claude Leny, sighed with relief: 'The French nuclear industry found "a new reason to live".'[18] The entire Lingao project was estimated to cost 18 billion francs (about US$3.4 billion) at the time when Framatome almost went bankrupt because there were no orders for NPPs. The project, said another senior manager at Framatome, would not 'fill our workshop, but without it our situation would have been extremely serious.' It would 'assure the equivalent of 9000 jobs per year, or an estimated 15 million hours of direct work.'[19]

The arrangement for the Lingao project was similar to that of Daya Bay: Framatome would supply 2x900MW nuclear islands based on the Daya

Bay PWRs and GEC Alsthom from the UK would supply two conventional islands, with EDF providing engineering support. The actual construction and assembly work would be done by Chinese enterprises under Framatome's and GEC Alsthom's supervision and with their technical assistance. A consortium of French banks would provide financing. The Chinese counterparts included CGNPC, Guangdong provincial utility, the CNNC and the Ministry of Electric Power.

Decisions on the Lingao project were controversial for several reasons. (a) whether another large nuclear power station was necessary around the same area in Guangdong; the province continued experiencing severe electricity shortages, but it would take another seven or eight years before it could generate electricity and high economic growth depended on the immediate addition of generation capacities; (b) whether the central government could finance such a large capital-intensive project at a time when the ratio of its revenue to GDP had dropped dangerously low, and when the China Light decided not to participate in another nuclear project (see Chapter 5) and (c) whether the country should import another turnkey nuclear power plant rather than reserving the resources to support its own nuclear industry (see Chapter 6). The Daya Bay project was a successful application of Western technology and enterprise practices; it started cultivating the nuclear safety culture, which was the necessary component for a sustainable nuclear energy industry. Yet, when the nuclear sector had to close down facilities and lay off its employees, giving such a large project to foreign companies was argued to destroy China's own industries (see Chapter 4). Many veterans in the nuclear sector were particularly angry about the 'betrayal' by the government for turning its back on an industry that had made the country so proud.

2000s: Rushing into nuclear energy expansion

After more than 25 years of continuing economic growth, by the beginning of the 21st century China had changed from being a minor and largely self-sufficient energy consumer to the world's second-largest and fastest-growing energy consumer, a major player in the global energy market and the second-largest polluter. To achieve the formidable goals set in the 11th Five-Year Plan – to double 2000 GDP by 2010 while reducing energy intensity by 20% – the power industry faced the most challenges. Several sets of challenges became more acute in the 21st century:

Supply/demand imbalances – it is well acknowledged that 'China has alleviated energy poverty on a scale and at a pace seen nowhere else ... all but around 10 million households now have some access to electricity.'[20] This development, however, has gone through a boom–bust cycle since the reform started: the country suffered power shortages during the periods 1982–90, 1993–97 and 2002–05. At the end of each period, there were signs of overcapacity, followed by an adjustment of investment. For example, at the end of the 1990s, as demand for electricity dropped, overcapacity reached 10%. Consequently, the government shifted its own investment to transmission and distribution networks in 1999 and ordered that no more investment would be approved for new power generation plants.

In 2002, the country started suffering from unprecedented power shortages that lasted until 2005. Of China's 31 provinces and major municipalities, 25 sustained significant power shortages. Industries suffered consequential losses and even experienced closures; households felt the impact of reductions in basic comfort levels. This was partly because economic growth picked up speed in the early 2000s after the country came out of the shadow of the 1997 Asian financial crisis. One indication of this growth was rising electricity demand, which jumped from 1253TWh in 2000 to 2288TWh in 2005, representing an average annual growth rate of 12.8%, exceeding the annual economic growth rate. Since 2006, the power industry has moved back to the boom part of the cycle, with an unsustainable average annual growth rate of 18.4%. The challenge today is how to minimise the fluctuations while expanding the electricity supply.

Another more urgent challenge is to meet the steady increase in electricity demand caused by urbanisation. 'Since 1992, nearly 200 million Chinese have shifted from rural to urban life, and the current pace of migration of about 15 million people per year moving into the cities is likely to continue for another 15 to 20 years.'[21] In China, energy consumption per capita of urban citizens is 3.5 times that of rural citizens and the electricity consumption per capita for urban population is more than doubled that of the rural population. This greatest migratory flow in human history has placed significant pressure on expanding electricity generation capacity as people switch from biomass and coal to modern energy, electricity and heat. The total installed power generation capacity in China has already become the second largest in the world (624GW in 2007), just behind the US (1088GW in 2007), and it is expected to double by 2020, reaching 1500GW.[22] This rapid increase in energy demand is putting enormous pressure on natural resources and the environment.

Energy security – given that about 80% of electricity is generated at coal-fired thermal plants, it is not difficult to imagine a situation where coal reserves will be depleted in the near future. No one wants to contemplate the prospect, not the Chinese government nor any international organisations or private companies; at least not openly for fear of triggering market panics. Yet, this is a real probability. China may have large coal reserves (about 12.8% of the world's total), but they are depleting quickly as production has been far exceeding its share of coal reserves. According to calculations of British Petroleum (BP), coal reserves in China will last no more than 40 years at the current rate of production.

Some Chinese scientists in the mid-1980s had already warned that the coal consumption would reach its peak of 2 billion tonnes a year by 2020 and

Table 3.2   Coal production in China (million tonnes)

	1981	1985	1990	1995	2000	2001	2002	2003	2004	2005	2006	2007	2008
China	616	872	1080	1361	1299	1382	1455	1722	1992	2206	2373	2526	2782
World	3381	4420	4719	4593	4607	4819	4853	5189	5588	5896	6189	6421	6781
% of World	16.1	19.7	22.9	29.6	28.2	28.7	30.0	33.2	35.7	37.4	38.3	39.3	41.0

Source: BP, Statistic Review of World Energy, Full Report 2009.

then would drop sharply.[23] China had already reached this level in 2004 (see Table 3.2). Recently, a group of academics at Uppsala University in Sweden confirmed this early study: at least 30–40% of the coal reserves in China have already been produced and the peak will be reached in 2020 or even earlier, and then there will be a rapid decline with little tail production. Energy insiders in China also acknowledge this situation, even though it is not an official line. As current Indian Prime Minister Manmohan Singh commented on the food situation in India, 'If India cannot produce enough food to feed the Indians, no one can'; likewise China needs to find a way to meet its own energy demands. Securing the energy needed to power the future is crucial and nuclear power becomes an increasingly attractive option.

Rising pollution levels – burning coal to generate electricity has been the most important contributor to GHG emission in China. China's power sector is the single largest culprit, responsible for an estimated 50% of the country's SO_2 emissions, 80% of NO_x emissions and 49% of CO_2 emissions in the mid-2000s. In addition to air pollution, thermal power generation (almost 80% of the total capacity) requires a large amount of processed water and contributes to severe water shortages in many parts of the country. They also discharge more than 70 million tonnes of solid waste a year.

According to a joint study carried out by the China State Environmental Protection Administration and the World Bank, the cost of air and water pollution adjusted to human capital ran at about 5.78% of GDP.[24] This does not include social and political costs. Air, water and soil pollutions have led to popular protests, and non-governmental organisations (NGOs) and sections of the media have lodged some of the most serious complaints against environmental pollution, which have now become political issues. China is also under increasing international pressure to reduce its CO_2 and other GHG emissions.

Two fundamental ways of dealing with climate change are by improving energy efficiency and increasing the share of renewable energy. China has promised to increase the share of renewable energy to about 15% of its total energy consumption. It is already the largest producer of hydro-electricity in the world. By 2008, with 170GW capacity, hydro stations provided about 14% of the total electricity. The country may have more hydro potential (estimated 540GW), but its development has slowed down considerably because of the problems of resettlement and environmental issues. It had built about 16GW wind power capacity by the end of 2009. With its low availability factor (15–20%), wind power is experiencing serious problems of getting on grids and having sufficient back-up capacities. It is also difficult to build large wind power mills at places where there are end-users.

Humans have always used wind, sun and rain as sources of energy – from windmills to dams. Nonetheless, wind, sun and rain suffer from intermittency. While the load factor (utilisation rate) for coal-fired thermal generation power plants is about 70–90% and that for nuclear power plants is 80–90%, for wind power it is about 30% and only 15% for solar. A coal-fired or nuclear generation plant with 600–1000MW capacity occupies a few square kilometres of ground space, while it takes several hundred square kilometres to build solar or wind generators with the same capacity. Even with advanced technologies to improve the availability of solar power from the current 15% to 36%, to generate one megawatt of electricity per year would require 40 acres of photovoltaic cells. Given the low availability rate, wind and solar powers need back-up or supplementary generators or battery back-up and this seriously limits their capacity as energy sources to combat CO_2 emission. Given that nuclear power plants can produce more than 1000MW at one plant, when compared with renewable energy, rated at several megawatts each, it would be a daunting challenge to expand renewable energy sources to meaningful levels in a short time.

Energy efficiency – after a steady improvement in energy efficiency for two decades (there was a 65% decline in energy intensity in 1980–2000), the trend of GDP growing at a faster rate than energy consumption reversed after 2002. Rising energy intensity (energy use per unit of GDP) threatens both energy security and the environment. Energy efficiency in various sectors in China is about 20–40% lower than in developed countries. There is clearly room to improve. Given that the power sector contributes almost half of CO_2 emission in the country, improving the performance of power plants, especially thermal plants, is a crucial component in dealing with the twin problems in China – energy security and climate change.

One way to improve energy efficiency and reduce GHG emissions is to close down small-sized (300MW and below), inefficient and highly polluting coal-fired generation plants and to replace them with the most advanced technologies in super-critical power capacity (600MW plus). In 2006, coal-fired generation units smaller than 135MW still accounted for approximately 30% of the country's total installed capacity. These smaller units consumed about 40% more coal than large ones to generate each kilowatt of electricity. Closing down small power plants, however, has been politically difficult. It often affects the very people and regions that need help and cannot afford to build large and more efficient plants.

To meet local electricity demand and meet the energy-efficiency target handed down from the central government, provincial governments have chosen the nuclear option to add more clean electricity supplies rather than closing down small dirty coal power stations. To build nuclear power stations would bring in the central government's investment, create jobs and build local infrastructure, from roads, water and sewage to electricity distribution networks.

These challenges require 'the accommodation of difficult-to-reconcile objectives: adequate energy for long-term economic growth, energy that can be secured without exposure to undue geopolitical risk, energy supply and utilisation consistent with long-term public health, and energy supply flexible enough to meet rising popular expectations for public and private goods.'[25] The Chinese government has decided to pursue an 'active' nuclear energy programme to supplement other measures, such as improvement in energy efficiency and development of renewable energy, in dealing with these challenges.

'Nuclear energy, as a proven, clean, safe, competitive technology, will make an increasing contribution to the sustainable development of humankind throughout the 21st century and beyond,' declared the minister of Industry and Information Technology of China.[26] In late 2002, the SPC submitted a report to the State Council, announceing that the country would build a total generation capacity of 800GW by 2020 and 4% of this would be nuclear generation capacity (an equivalent of 32GWe). The target was revised by the chairman of the China Atomic Energy Authority in 2004 to 36GWe by 2020. Two months later, Zen Peiyan, vice-premier of the State Council informed the Shanghai Institute of Nuclear Research that by 2020 China would build a total nuclear generation capacity of 40GWe.[27] On 2 March 2005, premier Wen Jiabao adjusted the government's policy on NPP development from 'appropriate' to 'active' development. This idea of 'active' development of nuclear energy was incorporated into the 11th FYP (2006–10) and then into the Medium- to Long-Term Nuclear Energy Development (2005–20) in 2006.

In the Medium- to Long-Term Nuclear Energy Development Plan, the NDRC provided a tentative plan for nuclear development (see Table 3.3):

Table 3.3   Tentative plan for nuclear development (GWe)

	Initial construction	Under construction	Continuation	Total capacity in operation
Prior to 2000				2.268
2001–05	3.46	4.68	5.58	6.948
2006–10	12.44	5.58	12.44	12.528
2011–15	20	12.44	20	24.968
2016–20	18	20	18	44.968

Source: NDRC, 核电中长期发展规划 (2005–10), October 2007, p.8.(NDRC, 'The Long- and Mid-term Nuclear Energy Development, 2005–10', October 2007, p.8.

The plan listed 13 sites in 7 provinces and all along the coastline. Interior provinces, especially those with rapid economic growth but without natural resources, such as Hunan, Jiangxi and Anhui, were also pushing to build nuclear power stations. High demand for new nuclear energy projects highlighted some perennial challenges the industry has faced – intensive capital costs, long construction periods, the need for intensive human capital, and safety and security concerns.

The high demand has also intensified some unique debates in China – who should be in charge of the projects, whose technology should be adopted, how can the interests of the general public and those of people who have to be relocated and resettled for nuclear projects be balanced, and how can the short-term and long-term environmental impacts of energy production and consumption be assessed and balanced? These are questions that have dominated the debate in nuclear energy development in China for the past 30 years and will continue to do so. Most importantly, the institutional and technical capacity or, rather more accurately, the lack of it, is the greatest challenge China's leadership faces in achieving the target of nuclear energy expansion. As one Western diplomat commented:

> The Chinese have competing government agencies in regulating nuclear power. All these entities gave a different opinion and a different vested interest in promoting nuclear power or putting it on the back burner.[28]

Uncertainty is the biggest enemy.

The next four chapters will examine the institutional configuration of agencies, choices and contingencies across four aspects of nuclear development – economics, technology, fuel services and environment.

Notes

1. '一九九五年国务院领导同志对核工业的重要讲话', 核经济研究, 6 (1995), 'Speeches on Nuclear Energy Industry from the State Council's Leaders in 1995', *Research on Nuclear Economics*, 6, 1995, p.3.
2. Yang, 2004, p.19; Barry Naughton, 'Zhu Rongji: The Twilight of a Brilliant Career', *China Leadership Monitor*, Winter 2002.
3. Ray Silver, 'Canadian and Chinese Provinces Agree to Study CANDUS in China', *Nucleonics Weekly*, 29:45, 10 November 1988, p.4.
4. Ray Silver, 'Canadian Team Hopeful for Chinese Project Breakthrough', *Nucleonics Weekly*, 30:13, 30 March 1989, p.5.
5. Ann MacLachlan, 'Engineering Advances for Next PWRs at Qinshan', *Nucleonics Weekly*, 36:7, 27 April 1995, p.8.
6. Ray Silver, 'China-Canada CANDU Deal Wobbles with Parties $1.8 billion Apart', *Nucleonics Weekly*, 37:17 25 April 1996, p.3.
7. Ray Silver, 'Canada, China Move Closer to Deal for two CANDU Reactors at Qinshan', *Nucleonics Weekly*, 37:29, 18 July 1996, p.4.

8. Ann MacLachlan, 'Daya Bay Units Inaugurated as China Firms Nuclear Planning', *Nucleonics Weekly*, 35:7, 17 February 1994, p.3.
9. '一九九五年国务院领导同志对核工业的重要讲话', 核经济研究, 6, 1995, p.3.
10. Mark Hibbs and Muelhein an der Ruhr, 'Framatome Deal Might Counter KWU's 41% Nuclear Contract Plunge', *Nucleonics Weekly*, 30:9, 2 March 1989, p.7.
11. Ibid.
12. Mark Hibbs, 'Sino-Russian Reactor Deal Signed', *Nucleonics Weekly*, 33:52, 31 December 1992, p.1.
13. Tai Ming Cheung, 'Power Politics: Chinese provinces pin hopes on nuclear plants', *Far Eastern Economic Review*, 6 April 1989, p.84.
14. Zhangxin, 'Tianwan Nuclear Power Station: Cooperation between Russia and China', *China Nuclear Industry*, December 2007, pp.26–28.
15. 'Speeches of State Council's Leaders in 1995', *Nuclear Energy Research*, 6, 1995, pp.3–8.
16. Mark Hibbs, 'Russian Vendor Experts Predict Lingao-1 Won't Start Till 2004', *Nucleonics Weekly*, 36:49, 7 December 1995, p.12.
17. Mark Hibbs, 'China Will Abandon Liaoning as Site for VVER-1000 Project', *Nucleonics Weekly*, 37:42, 24 October 1996, p.8.
18. Ann MacLachlan, 'China Strikes Deal with French for Two New PWRs at Lingao Site', *Nucleonics Weekly*, 36:3, 19 January 1995, p.1.
19. Ibid.
20. IEA, 2007, p.281.
21. Kenneth Lieberthal, 'Challenge and Opportunities for US-China Cooperation on Climate Change', Testimony before the US Senate Committee on Foreign Relations, 4 June 2009, p.7.
22. These are the data provided by the US Energy Information Administration (EIA). The one provided by the Chinese sources indicates the total installed generation capacity is about 700GW in 2008.
23. Guo Xingqu, 'Role of Nuclear Power in China's Future Analysed', *Journal of Dialectics of Nature* (in Chinese), 2, 10 April 1986, pp.25–35, in 'China Report: Economic Affairs', JPRS-CEA-86–093, 7 August 1986, p.89.
24. World Bank and SEPA 2007.
25. Steinfeld, 2008, p.133.
26. Li Yizhong, 'Concluding Statement by the President of the Conference', at *IAEA Ministerial Conference on Nuclear Energy in the 21st Century*, Beijing, April 2009.
27. 定军, 2005.
28. Kevin Platt, 'China's Nuclear Power Program Losses Steam', *The Christian Science Monitor*, 21 July 2000.

4
Who Decides? The Politics of Nuclear Energy

A popular image of China is: 'a large unitary state characterised by an unusual degree of cultural and linguistic homogeneity, a tradition of statehood that stretches back into distant antiquity, and a government that insists on representing itself in strict post-Westphalian terms as sovereign, unitary and rational.'[1] Many observers of China still hold that: 'China remains authoritarian'[2] because of the Communist Party's 'exclusive guidance of economic, social, military and political goals.'[3] The Party's leadership is guaranteed institutionally by the boundless power of the Politburo and its Standing Committee. The government is no more than a servant of the Party.

This image of China as 'a one-party state that would think, speak, and act with one mind, one voice, and one purpose' evaporates quickly as soon as one investigates any policy issue.[4] Those who have done extensive studies on energy policy in China, in particular, have come to the conclusion that 'energy policy in China today is a battleground of negotiation among powerful actors with conflicting interests that are evident at all levels of analysis.'[5] Indeed this energy institutional landscape, characterised by overlapping jurisdictions and inconsistent waves of centralisation and decentralisation, is not new.

Examining the decision-making processes over energy policies about 30 years ago, Lieberthal and Oksenberg concluded that within the Chinese authoritarian system there was a scattered, disconnected and layered governmental structure that required any policy initiatives gain the active cooperation and support of many separate and competing bureaucratic units that effectively have mutual veto power. 'Policy is the aggregate response of leaders or factions of the participants, their strategies for advancing their beliefs and political interests, and their differentiated understanding of the problem at hand.'[6]

To understand China's fragmented authoritarian decision-making process, some scholars focus on the roles and controls of individual leaders, especially their competing interests and struggle for leadership.[7] Some focus on the institutions, particularly the lack of centralised institutions and capacities in coordinating policy making.[8] Some emphasise the importance of informal networks in the fragmented decision-making processes. Others argue that how decision making is fragmented varies from one policy issue to another.

Nuclear policy is one of the issue areas where we would expect a relatively closed policy-making process and consequently much more coherent and integrated policies. This was the experience in most countries with nuclear energy programmes. In France, a tight network of bureaucrats and engineers 'wove links between technology and national identity into the fabric of reactor designs and program development.'[9] In South Korea, it was the government that created institutions, decided the national strategies, fostered industries and eventually invested in nuclear development. There was not only a coherent national strategy but also a set of institutions that translated the strategy into real development. The public were not involved until long after the nuclear programme was put in place. Similar closed policy-making circles on nuclear development can be easily found in Britain, Japan and many other countries.

In China, however, there has never been a set of coherent nuclear energy policy, nor has there been a centralised government agency in charge of making nuclear energy policies. This is the result of several developments:

First, nuclear energy came out of the weapons programme – for any new policy issues to appear on the agenda, they need powerful political champions to change the expectations and the existing rules of the game. Major public policies, scholars have long argued, 'constitute important rules of the game, influencing the allocation of economic and political resources, modifying the costs and benefits associated with alternative political strategies, and consequently altering ensuing political development.'[10]

Second, the economic reform in China has been accompanied by several rounds of government reorganisation. Institutions responsible for various energy sectors were eliminated, merged, recreated or reorganised at frequent intervals. For example, two rounds of reorganisations were conducted in 1982 and 1988 to make 'the state' more compatible to the economic reform, In 1982, the ministry in charge of the nuclear weapons programme was renamed the Ministry of Nuclear Industry (MNI) to reflect the government's desire to move into the civilian nuclear programme. In 1988, the MNI was merged with the rest of the energy sectors to the Ministry of Energy (a) to downsize the government and (b) to separate government from economic functions. The first objective might have been achieved in terms of numbers but the second object was quickly aborted.

In the 1990s, government reorganisations were conducted to refine some functions of the government in the national economy. In 1993 and 1998, the number of institutions under the State Council shrank significantly as the ministries responsible for industries were abolished while their business segments were commercialised and corporatised. In this process, the nuclear industry lost its status as a ministry but failed to be corporatised successfully. No one government agency was in charge of its policy making while all wanted to maximise their interests if opportunities opened. Nuclear energy development lurched forward with several projects yet without a long-term strategy or a well-thought-out national plan.

Third, when nuclear development found its home in energy, it fell into the notorious 'fragmented energy policymaking structure' with, as a Chinese official describes, its 'weak coordinating capability, inadequate policy enforcement ability, insufficient social supervision, inconsistent central and local policies, substandard regulation entangled with loopholes, inadequate administrative and regulatory effort, and severe personnel shortages.'[11]

This chapter examines the political changes that the nuclear sector has gone through, and especially focuses its discussion on the idea of nuclear energy looking for high-level sponsors in Beijing, and the *ad hoc* nature of the first few NPPs. Understanding nuclear energy development in China, therefore, becomes a task of identifying and understanding the interests of individual top leaders and those of ministries, their roles, resources, arguments and specifics of their pertinent functions in the policy-making process.

Nuclear energy: An idea seeking a champion

The world's nuclear industry is organised in two ways: one emphasising its 'electricity' component, and the other the 'atom' component. In countries, such as Brazil, Britain, France, Japan, Mexico, South Korea and the US, the public utility companies (some are public while others are private) invest, construct, operate and manage NPPs, while research and design are carried out by the government atomic agency. In countries such as Argentina, India, Pakistan and the former Soviet republics, the nuclear energy programme is an integral part of the atomic sector under the direct control of a government agency, independent from public utilities. In both organisational models, public institutions serve as instruments of national policy, mobilising public support, providing economies of scale, sponsoring mission-oriented research and development (R&D), coping with the scale of risks and benefits, and taking the long-term view. Safety, waste management and uranium enrichment are managed and regulated by separate and independent institutions.

Typically, 'government-owned national champions were used to develop a range of industries, especially the commanding heights of the national economy – energy and electricity supply.'[12] Government agencies in charge of R&D have established close ties with the companies that build and operate NPPs, as in the case of the British Atomic Energy Authority and the Central Electricity Board or in France, CEA and EDF.[13] Even in the US where private ownership is the rule, the close tie between the government-sponsored

R&D and utilities has been at the core of its nuclear energy development. The centre stage the government occupies in the nuclear industry is necessary because of the economic, technical and safety nature of the industry. Governments' long-term strategies are fundamental for the success of nuclear energy development and expansion.

In China, the nuclear industry started as a weapons programme and consequently followed the model of the second group – it was 'atom' rather than 'power' that dominated the organisational structure, policies, resources allocations and development in general. When Deng Xiaoping started pushing his 'strategy of demilitarisation, liberalisation, and opening up of China's economy to market forces through closer integration with the international economy,'[14] the military-defence industries, nuclear included, lost their privileged status and priority access to national resources. To make radical shifts from 'atom' to 'power' and from serving the national military and defence to engaging in civilian production required organisational reconfiguration.

In 1982, the 2nd Ministry was renamed the Ministry of Nuclear Industry (MNI) to facilitate the process of moving into the civilian nuclear programme. Along with its counterparts in the defence-military industries, the nuclear sector was supposed to adjust and adapt to the new environment and general economic reform by 'combining the military and civil, combining peace and war, giving priority to military products, and making the civil support the military.'[15] If a nuclear energy programme were the future for the nuclear sector, MNI would have to depend on Li Peng, who was the deputy minister of energy, 'to be its voice among the 25 to 30 top policy makers, and those 25 to 30 in turn looked to Li as their links to the energy bureaucracy.'[16] Nuclear energy, however, was not seen as the future for the nuclear sector nor an option for its conversion or for its reinvention. In organisational terms, instead of looking for political patrons in the energy arena, MNI had their patrons in the military and military-defence, who soon lost their political favour of Deng.

There are several reasons the people in the nuclear sector did not see nuclear as their future or understand the necessity of switching their loyalty from the weapons programme to a nuclear energy programme.

First, nuclear energy development was at its embryonic stage and it would take a long time and huge resources for its development to get off the ground, the resources that the nuclear industry could not spare. No political leaders would be likely to make a political issue or a career out of an infant industry with high risks. Furthermore, nuclear energy development would be too expensive and too slow to attract people who sought to take advantage of restructuring the military and defence industries. In August 1982, Hu Yaobang told the minister of Nuclear Industry that he should take full responsibility for the slow nuclear energy development and the Ministry should support and participate in nuclear energy development.[17] As far as the MNI was concerned, however, even if a nuclear energy programme could be developed quickly it would not be able to solve the problems it faced – substantial budget cuts from the central government, significantly reduced military orders, excessive numbers of employees, narrowly focused specialties, remote locations and a heavy social burden.[18]

While the majority of the Ministry's thousands of factories, research institutes and other facilities were located in remote areas, potential nuclear power stations would only be built along the coastline. At the time, relocation was strictly regulated and it was almost impossible to move from interior remote places to cities. Even if nuclear energy projects could create jobs for a few hundred of its technicians and skilled workers, there were no other employment opportunities for their families and the next generation, which was part of the responsibility of MNI. With more than 200 enterprises and 300 000 employees, covering the construction and operation of nuclear facilities, the prospecting and mining of nuclear and related fuels, and the disposal of waste, under its flag, the MNI would have to find new opportunities to employ these people and take care of their basic needs. Nuclear energy was just not a valid option.

Second, even though the traditional nuclear programme was a military programme, under the direct control of the Party Politburo, it was subject to frequent juggling of competing interests, internal debates and fights for what the priorities should be and where resources should go and consequently organisational changes.[19] The three main institutions representing the interests of political leaders, military planners and scientists seldom agreed on any specific policies regarding China's nuclear programme, military or civilian – these institutions were the Ministry of National Defence under the control of the Party Central Military Commission, headed by Mao and Later Deng, the Defence Science and Technology Commission, in charge directly of the 2nd Ministry, and the Institute of Atomic Energy under the Chinese Academy of Sciences. Ministry of National Defence and the 2nd Ministry preferred an expansion of nuclear weapons programme, while the Institute of Atomic Energy demanded more resources for nuclear energy development. This struggle among the competing interests might be the most important reason that by the 1970s China was the only 'nuclear state' without nuclear power.[20] After MNI was established, while the competing interests could not agree on where the industry should go and how it could reinvent itself, they seemed to agree that nuclear energy was not an option that could save them.

Third, various policy options were discussed and implemented to adapt and adjust to the general reform, but nuclear energy development was not considered by those in MNI.[21] They tried to get into the production of consumer goods, such as televisions, refrigerators and electric fans, but there was no advantage in doing so. MNI borrowed heavily on behalf of some units so that they could get into the petrochemical industry. People were even sent to the Daqing oil field where they were resented by the employees in the petrochemical industry for encroaching into their territory. They failed and ended up with more losses than gains. Among the civilian goods produced by the nuclear industry, isotope irradiation devices were the main money-making products.

Some proposed to export its surplus of enriched uranium and 'some nuclear related products abroad'. Some suggested China should accept nuclear waste for storage from other developed countries.[22] Both were designed to make money and especially in foreign currency earnings, but neither of the ideas really worked. The nuclear sector 'failed to create reliable, "main-stay" product lines or develop a more consumer-savvy attitude when it came to price, quality, and adding new features.'[23] Yet, nuclear energy did not enter discussions even though the State Council repeated its message, first sent out in 1980, that the nuclear energy programme was now under the jurisdiction of the MNI and the industry should take advantage of the opportunity to adjust itself to the new environment. In 1983 and 1984, Zhao Ziyang, the Premier of the State Council, specifically told the MNI that it should start R&D on nuclear energy not only for electricity but also for heating, the technology that was considered to be advanced even in developed countries. Another project was approved for MNI to work together with Tsinghua University on high-temperature gas-cooled-reactor technology. Tsinghau University went ahead with the project, but MNI showed little interest because the central government had allocated limited resources to the programme due to its own shrinking revenues.[24]

Fourth, a 'planned economy mentality'[25] and a mentality of superiority of many in the nuclear sector did not help the industry in its conversion or to see nuclear energy as an option, as it was explained by the minister of Nuclear Industry:

> The nuclear industry in the past was solely concerned with war production and operated as an independent system. Its assignments were handed down from the higher levels. Its products were sold by the state on a quota basis. Special fund allocations were made available for it. Its material and equipments were guaranteed on a priority basis. The switchover to civilian use has called for its operating on its own and making independent exploratory and operational research efforts and for its meeting market and consumer needs with regard to economic, technical, quality and service levels.[26]

With this mentality of entitlement, many resented the central government for marginalising and abandoning an important sector. They insisted that they deserve the continuing support of the central government because the industry had made a great contribution to the country's nuclear weapon and missile programme. Their attitudes did not help discover new opportunities and develop innovative ideas.

This mentality of being privileged was to an extent supported by some on the top. Zhang Aiping, vice chairman of the Central Military Commission of the CCP and long-time aide to Nie Rongzheng, was responsible for overseeing and coordinating defence science, technology and industrial affairs, which put him in effective charge of the defence economy. He repeatedly defended the nuclear sector, arguing for continuing and increased government support.[27]

In the mid-1980s, Deng increasingly considered these senior military and defence officials, who had defended the record of the nuclear sector and demanded more assistance to save the country's greatest industry, an impediment to the reform. Deng wanted a nuclear industry that would transform itself and be able to support itself when the central government was no longer able to bankroll the loss-making industry the way it used to. As he grew impatient with some senior military and military-defence officials, referring to them as 'undisciplined, arrogant, extravagant and lazy', Deng initiated 'a series of deftly engineered personnel shuffles, including a wholesale pruning of the PLA's bloated senior officer corps' to neutralise the military and military-defence industries to his reforms.[28] Those who did not retire were asked to retreat from the Central Committee of the CCP to the National Committee of the People's Political Consultative Conference – a nominal adviser to the government.

Deng's decision to demobilise one million soldiers in 1985 led to further cuts in military budgets and military orders for the nuclear sector. Despite all the pressures and losing a political protector in a high place, the industry as a whole did not accept nuclear energy development as a viable alternative for its survival and revival.[29]

Those who did advocate nuclear energy development and wanted the nuclear sector to take the lead were almost exclusively nuclear scientists, both military and civilian. They were not in government and spoke with authority only because of their expertise. They advised that nuclear energy could be the future not only for country's future energy needs but also for China's nuclear industry. Yet, by the time their advice was accepted by some senior officials in the sector, it was too late – the industry had lost its political status, its spokesmen in high places, and many talented people as well. More importantly, China's nuclear energy development had already gone on a route driven by local and specific interests rather than a national policy.

In 1987, the MNI provided its vision of the future for nuclear energy: 'the Minister of Nuclear Industry should become a "second energy minister" to do a good job in the peaceful use of nuclear energy.'[30] The Minister's vision was crushed a year later when the central government dissolved the ministries of coal, petroleum, and the nuclear industry, split MWREP and created the Ministry of Energy (MOE). The nuclear industry was placed under the Ministry of Energy, together with the power industry, the electric machinery portion of the Ministry of Machinery Building Industries, the coal and petroleum industry. MOE was designed to 'comprehensively regulate all energy industries in the country, to be responsible for mapping out principles, policies and strategic development for the energy sector, to work for overall balance and macro-decision making, promote rational utilisation and development of energy, formulate relevant laws, regulations and policies, supervise and coordinate production and construction, improve economic results, formulate technology policies and promote energy conservation and overall utilisation of energy by the whole society.'[31]

The reorganisation of 1988 was an effort by the central government to cut the size of the State Council, improve its efficiency and eliminate bureaucratic red tape. It was also designed to centralise energy policy-making and the control of investment allocation in all energy sectors. 'Unfortunately, the MOE was little more than a collection of the same vested interests within one umbrella organisation, the same personnel, the same allegiance, and the same entrenched interests.'[32]

Each energy sub-sector had its own problems and agenda, and maintained its own political patron. The coal industry, for example, was a perennial loss-making industry, which by 1992 had acquired a total loss of 5.75 billion yuan.[33] It was also an industry where local interests exerted significant influence. The petroleum industries that had amassed a huge amount of debt, which forced the industry to face the prospect of insolvency, soon found their political support in the form of the new government under Zhu Rongji. Zhu was willing to give a great deal more financial and administrative autonomy to the large state-owned corporations such as the petroleum industry, on which the Chinese Government had developed fiscal dependence, to prepare for the listing of their subsidiaries on domestic and international stock exchanges.[34]

The nuclear energy did not find its place under a single umbrella either, for several reasons. First, as an insider explained, in the reorganisation, even though the MOE 'struggled vigorously for leadership in nuclear power development', the State Council preferred to have direct control over civilian nuclear development through the leading group headed by Li Peng,[35] who became the Premier of the State Council in 1988. Second, regarding the policy of separating the government's 'main economic responsibility from direct management of economic enterprises to macro supervision and regulation'[36], the government functions of MNI were absorbed by the MOE. The economic and management responsibilities were taken over by a newly created state-owned company – the China Nuclear Power Company (CNNC). The nuclear sector under CNNC continued its struggle for identity as it was accountable to both the State Council and the COSTIND, along with the rest of the military-defence sectors – aviation, space/missile, ordnance and shipbuilding. Given that COSTIND controlled 'only research, development, and production of certain high technology weapons and provided related policy guidance,'[37] CNNC was seen as part of the military-defence industry, taking little interest in nuclear energy.

Third, under the CNNC banner, there were more than 200 enterprises, covering the construction and operation of nuclear power plants, prospecting and mining of nuclear and related fuels, and the disposal of nuclear waste. Instead of becoming productive forces to 'help feed mainland China's civilian sector...and expand into new and profitable areas',[38] most of CNNC's 300 000 employees struggled to survive in their military and semi-military segments and were constantly begging for more support and subsidies.

The Ministry of Energy was never in a position to take over nuclear energy development, mainly because it was paralysed with internal conflicts and by 'a collection of bickering interest groups.'[39] Five years later, the Ministry of Energy was dissolved and several line ministries were recreated – the Ministry of Coal and the Ministry of Electric Power (MEP), but not the Ministry of Nuclear Industry. The nuclear industry remained under the umbrella of CNNC as a company that was running heavy losses. Nuclear energy had missed another chance.

Many in the sector blamed their downgraded status as the main reason for its downfall and repeatedly called for the government to restore the Ministry of Nuclear Industry to secure direct access to the central government's budget allocation and decision-making process.[40] Others started to realise that concentrating on civilian nuclear programmes was a way out; 'without economic wealth, it did not matter whether you enjoyed the status of ministry or direct subordination of the State Council.'[41] By then it was too late; the industry was trapped in a Catch-22 situation. Expanding the nuclear energy programme seemed to be the main option for the sector to revive itself, but Zhu Rongji made it clear that the government would not approve any new nuclear energy projects until CNNC became profitable. By then, many of its factories had closed, many talented people had left the field, especially those that the future nuclear energy development would depend on, and many more were 'laid off' with minimum support as part of the larger restructuring of SOEs. The industry was completely 'demoralised' in 1999 when, under Zhu, another round of government reorganisation forced the CNNC to be split into two – the China Nuclear Energy Corporation (CNNC) and China Nuclear Engineering and Construction Corporation. Most of the loss-making segments, such as uranium exploration and mining, were decentralised to provinces, but CNNC had to find a way to finance the transfer and lay-offs.[42] It was not until 2003 that the nuclear sector was finally able to turn around from a loss-making venture to a profit-making venture, mainly thanks to its nuclear energy projects.

Just as the nuclear sector did not see nuclear energy as its future, the power sector did not see nuclear energy as the future for the country's electricity. In contrast to the struggling nuclear sector in the first two decades of the reform, the power sector had built on its success because, as the World Bank officials summarised:

> First, the government has been able to set and focus on priorities at each stage rather than diffuse limited institutional capacity implementing reforms on multiple fronts. Second, policy makers have been able to pilot different approaches in a few provinces, await preliminary results, determine a preferred strategy, and then pass the enabling legislation required to mainstream the selected strategy. Third, the approach has made it possible to build broad consensus by acknowledging regional diversity in the power sector and trying alternative approaches to achieve the same objective. Fourth, and perhaps most important, the government has been *consistent* and *determined* in moving the power sector toward a commercially stable future.[43]

Consistent and coherent policies were not possible without political champions and people speaking up for nuclear energy in high places. From the beginning of the reform, the power sector was led and later protected by Li Peng, a Soviet-trained electrical engineer, who was not 'an old revolutionary...nonetheless was associated with the communist movement almost literally from birth.'[44] As Li Peng moved up the political ladder, he took his interests in electricity along with him to the core decision-making circle. He argued and advocated that in building up primary industry, the country must first speed up energy construction, centred on electric power. This belief was pushed forward, with new ideas and concepts on how to bring about resources for electricity expansion, which included lowering entry barriers to encourage investment made by local government and large enterprises, and the utilisation of foreign capital, especially multilateral and bilateral economic assistance.[45] Li was able to move ahead not only 'because of his martyred father and illustrious patrons', but also because he 'exhibited good technical mastery over his field.'[46] Thus, nuclear energy development fell into his hands.

In 1983, the State Council formed its Leading Group on Nuclear Development, headed by Li Peng, vice-premier, and two deputies, the deputy chairman of the SPC and the deputy minister of Nuclear Energy. The Group also included members from the SPC, the SEC, the COSTIND, the Ministry of Foreign Affairs, MNI, the Ministry of Machine Building, MWREP and the Ministry of Foreign Trade. The Leading Group was designed to coordinate ministries and, in fact, became a forum for ministries to compete and compromise on their diverse interests. Some analysts argued that 'since the original task of the 2nd Ministry did not include the civil application of nuclear power...the Chinese Ministry of Power claimed its leading role in nuclear power.'[47]

There was an initial struggle for leadership when the nuclear energy issue was first discussed in late the 1970s. Yet, when the first two nuclear projects were underway, neither MNI nor the MWREP wanted them. Indeed, Li Peng, who had the political backing of Deng and had a vision on how to build a nuclear energy programme, at one time suggested MNI take over the nuclear energy programme because of the 'crowded agenda' of MWREP. MNI passed on the offer because of its internal disagreements on many issues.

The power sector wanted to put all the resources into expanding power generation capacity as soon as possible and nuclear energy was unable to achieve this goal. At the beginning of the 1980s, the MEP called for a gradual but eventually 'immense' increase in investment in power generation; the major emphases were on: the expansion of hydro capacity, accelerated construction of coal-fired mine-mouth thermal power plants and the speedy development of large generation plants along the coast.[48] The central government introduced a policy of multiple channels of investment in power generation that would allow third parties other than the central government – mainly provincial and local governments, but also domestic companies – to invest in power generation. It also decided to approach multilateral and regional financial institutions, such as the World Bank and Asian Development Bank (ADB) for assistance.

Loans from the World Bank were the main source of sovereign borrowing in the power industry. From 1980 to the end of 1992, the power industry signed contracts of US$9.7 billion in loans with the World Bank, the ADB and foreign governments, all of which were prohibited from financing nuclear power projects. Meanwhile, none of the domestic players was willing or able to put the required capital together quickly for nuclear power projects. Finally, the MEP simply never saw NPPs as a viable alternative to meet the demand for electricity.[49]

By the early 1990s when the reform resumed, the power sector was well on its way to expand the country's generation capacities in both thermal and hydro stations, large and small. In great contrast to the nuclear sector, the power sector was not only profitable but also expanding fast. When the Ministry of Energy was dissolved in 1993, the electricity industry managed to have its ministry restored. With the MEP in place, the sector was centralised in the process of commercialisation and corporatisation, which led to the creation of the State Power Corporation of China (SPCC). As a vertically integrated monopoly, SPCC co-existed with the MEP for two years before the Ministry was disbanded in 1998.[50] Nuclear energy remained outside the MEP and SPCC.

With two potential hosts of nuclear energy showing no interest in its development, and without a central champion for its development, nuclear energy projects were picked up by specific and local interests. The central government approved the Qinshan I and Daya Bay projects, both of which then took on a life of their own under two quite independent institutions – a small team centred on the 728 Institute and Guangdong province. For Tainwan and Qinshan III, little debate took place in the central government because they were more political and diplomatic exercises than economic or energy projects. A nuclear energy programme did not develop until the 2000s when both the nuclear and power sectors saw the opportunities in expansion and the nuclear sector was brought under the energy umbrella. Yet, the old structure and inertia prevented the nuclear industry from integrating into and competing with the electricity sector.

Fragmented decision-making

Entering the new millennium, China struggled to meet rising energy demands and to deal with environmental pollution as the result of its heavy reliance on coal. Governance became a more pressing issue too, as China's fragmented energy bureaucracy was one of the main contributors to the country's unsustainable energy situation. The old central planning in terms of organisational structure was long gone but with the notorious 'fragmented bureaucratic authority, decision making in which consensus building is central, and a policy process that is protracted, disjointed, and incremental' continued.[51] The finding of the study conducted by Liberthal and Oksenberg 30 years ago that 'the structure of the energy sector highlights the fragmentation of authority' remains a valid description of the policy-making today. This is a challenge the central government has tried to deal with since the early 2000s.

The last time China had a single centralised institution coordinating policy making for all energy sub-sectors, including nuclear, was in 1993 before the Ministry of Energy was disbanded. In 1998, the last two standing energy ministries, coal and electricity, were dissolved too. Their administrative functions were transferred to the State Economic and Trade Commission (SETC). With the rise of Zhu Rongji, the SETC became the 'mini State Council' while long-term and short-term planning in the energy sectors was significantly dispersed. The SPC survived the 1993 reorganisation, although it was apparently weakened as for the first time it was headed by a non-Politburo member. In theory, the SPC was still responsible for making national economic plans, approving major projects and setting prices for all energies. The function of formulating long-term economic plans, however, was shared with the State Commission for Restructuring the Economy, which was created in 1982 and was tasked with studying the interaction between economic, technical and foreign developments and China's political, social and economic reform.

The system was deemed unworkable and within a few years both macro-level commissions (SPC and SETC) were defunct. The SPC, whose name changed to the State Planning and Developing Commission in 1998, dropping the last vestige of the planning economy, was renamed the National Development and Reform Commission (NDRC) in 2003 after absorbing some functions of the SETC, which was eliminated. No longer did NDRC have to compete with its counterpart, SETC. Nonetheless, it faced powerful players in another category – large state-owned corporations. It became increasingly clear that the country should create a mega-ministry to be charged with the authority in making energy policies, coordinating the activities of all energy sectors and with other sectors of the economy, and ensuring policies be implemented according to their design.

Unlike most countries, there is no one ministry or department in charge of setting a broad energy strategy and making energy policies for the country as a whole, as the US Department of Energy (DOE) or the French PEON (production d'électricité d'origine nucléaire) – 'an influential advisory commission formed by government officials, EDF, CEA, and industry' in the 1950s.[52] In China, more than a dozen government agencies at the central level can claim to have authority over energy policy-making.

Among them, the NDRC sees itself as the ministry above all others. As a macroeconomic planning agency, the NDRC has many responsibilities, among which are: deciding macroeconomic policies, planning for the national economic development, approving investment, setting prices, ensuring agricultural and regional development, stockpiling strategic commodities and high-tech development. It has three relevant bureaus for energy:

- Energy Bureau, responsible for energy policy and strategy making and project approval.
- Pricing Department, responsible for setting prices for the energy sectors.
- Environment and Resources Comprehensive Utilisation Department, in charge mainly of energy efficiency.

Its energy bureau was supposed to work out the national energy strategy and coordinate the actions of various energy sub-sectors. Until 2008, however, it employed fewer than 50 people. It had to coordinate with other departments in project approval and energy price setting. This proved more problematic for nuclear energy because it did not even have the authority to approve nuclear power projects. That authority was in the hands of the State Council, the equivalent of the cabinet in the Westminster parliamentary system of government.

When energy shortages reached a critical stage from 2002 onwards, it was imperative that there be better coordination of all energy sub-sectors. In 2005, the State Council created the National Energy Leading Group as a new inter-agency body to improve policy making and policy coordination among fragmented energy sectors. The Leading Group was headed by Premier Wen Jiabao, with two vice-premiers serving as vice-chairs. It included: leaders from 13 major government agencies; the NDRC; the Ministry of Foreign Affairs; the Ministry of Commerce; the Commission of Science, Technology and Industry for National Defence; the Ministry of Defence and others. The group was responsible for formulating the country's energy strategy and for providing policy suggestions to the State Council regarding energy exploration, conservation, security and international cooperation within the energy sector. NDRC Minister Ma Kai led the 24-person, vice-ministry-level State Energy Office, charged with overseeing day-to-day work for the Leading Group. The idea of creating a single agency appeared to be sound in order to develop a coherent national energy strategy, monitor energy security, organise energy-related research and coordinate activities of all state-owned energy corporations. Nevertheless, the effort failed. The Leading Group met only twice in the next two and half years – 2 June 2005 and 20 April 2006. Little was achieved.

Leading groups had long been used to strengthen the domination of the Communist Party over government affairs, as was explained by the Central Committee of the CCP in the early 1950s when the first leading group was created:

> These leading groups belong to the party centre, are directly subordinate to and directly report to the Politburo of the CCP Central Committee and the Secretariat. Strategy and principal policies are dominated by the Politburo, and detailed arrangements are the responsibility of the Secretariat. There is only one, rather than two, 'political design institute'. All major strategic policies and concrete arrangements are centralised; there is no dispersion between the party and government.[53]

The Leading Group is the preferred mechanism to coordinate policies and works in a similar way as the 15-Member Special Group headed by Zhou Enlai in the early 1970s, the Nuclear Development Leading Group headed by Li Peng in 1983, or the National Energy Leading Group of 2005.[54] These leading groups are created by the State Council and can be *ad hoc* depending on the importance of the issue of the day and the difficulties of coordination among various ministries. For those leading groups of the Central Committee of the CCP, such as Finance and Economics, Taiwan Work, or Combating Commercial Bribery, ensuring party dominance of decision-making power on strategies and major principles is the first priority. For the State Council's leading groups, it is often more pressing to have the cooperation of various ministries. For example, in the 2000s, the State Council also created the Leading Group of the State Nuclear Power Self-Reliance Work to coordinate the strategy of introducing the most advanced nuclear technology and its adoption and absorption. Their effectiveness often depends on how urgent the matter is, how much attention the Premier decides to give it, and who takes the lead if the Premier does not.

The National Energy Leading Group, created in 2005, was undoubtedly formed to centralise the decision-making authority on all energy issues. This Group, however, was different from many others created since the reform because it was led by the Premier and was an initiative and a prerogative of the government. In the energy sector, even though all major players are state-owned corporations, they do not operate in the same way as old-style state-owned enterprises (SOEs) in a planned economy, nor old-fashioned nationalised companies run by the government and designed to control chunks of the national economy. Their pursuit of self-interests was the very reason for the need to create a centralised institution to coordinate policy-making and their activities. In the end, the Group failed to achieve a great deal, also because of the changed players. Often leading groups were created to 'provide a forum for bargaining and compromise, or rather consultation and reconciliation ... and for policy analysis and assessments to reach a consensus.'[55] None of the ministries represented in the Group could speak on behalf of all energy sub-sectors or had direct jurisdiction over all the major state-owned energy corporations.

The problems and ineffectiveness of the National Energy Leading Group quickly became apparent. 'Some Chinese and foreign commentators even maintained that China's energy security was undermined by the very institutions responsible for enhancing it.'[56] The central government held extensive consultations with energy SOEs, academics and think tanks, looking for suggestions on a new governance structure for energy sectors. In 2008, after nearly five years of intensive national debate, the National People's Congress (NPC) suggested the creation of the National Energy Commission (NEC) – a senior level discussion and coordinating body put together by the State Council – and the National Energy Administration (NEA), in charge of managing the country's energy industries, formulating energy strategies, drafting energy plans and policies, negotiating with international energy agencies and approving foreign energy investment. NEA replaced the Energy Department of the NDRC and absorbed the nuclear power administration of the Commission of Science, Technology and Industry for National Defence (COSTIND), which was downgraded again to a department, and placed under the Ministry of Industry and Information Technology (MIIT). This is the first time ever that nuclear has been brought under the umbrella of national energy policy-making.

NEA is not a 'super' Ministry of Energy, as many recommended and expected it to be. It can be seen as a halfway house between a full ministry and a subordinate department of the NDRC. It is set at the vice-ministerial level, accountable to both the NDRC and the State Council. The head of NEA is also the deputy minister of the NDRC, but NEA did not sever its relationship with the NDRC. 'One of the uncertainties surrounding NEA's establishment is how much autonomy it would have from the NDRC on energy policy.'[57] Indeed, it is not clear about the line of accountabilities: would it report directly to the State Council and the Premier or would it be accountable to the NDRC? To what extent can it deal with other relevant ministries on an equal level since it was set below the ministerial level? A year after its creation, the State Council announced the size of this organisation – 120 full-time positions. This is a minute number compared with energy ministries in other countries. The US Energy Department has more than 10 000 employees in its headquarters.

According to the deputy director of the newly created NEA, Sun Qin, the creation of NEA streamlined the decision making for nuclear energy. In the old days, the NDRC was in charge of planning, siting and project approval, as well as acceptance checks; COSTIND was in charge of technology development, personnel training; NSSA was in charge of safety, among other things. Although safety still belongs to an independent regulator, everything else in terms of strategy, policies, R&D, technology, equipment manufacturing and international cooperation was under the jurisdiction of NEA. Not everyone agreed with Sun Qin's analysis in 2008 and it is not clear he would agree with his own statement a year later after he took over as the general manager of CNNC in August 2009.

The organisation of the Chinese nuclear industry remains fragmented. The State Council sets the overall plan, the national strategies and specific development targets. Under the State Council, several ministerial level institutions are in charge of different aspects of nuclear energy: the NDRC and its NEA are in charge of project approval (but in practice all nuclear power projects must have the approval of the State Council); the Ministry of Environmental Protection (MEP) is in charge of environmental assessment and environmental approval for siting and other aspects of nuclear power station operations; the Ministry of Finance sets policy on taxation and lending for the policy banks; and the Ministry of Land and Resources sets policies and regulates uranium exploration and mining, and land use. These ministries all have some responsibilities over the nuclear energy industry, but they do not overlap and it is unlikely that these ministries would run into conflicts over specific policies or nuclear projects. Because of this, people in the industry always point their finger at the State Council and the NDRC.

According to many Western observers, however, 'nuclear energy policy is determined by the China Atomic Energy Authority with the approval of the Commission for Science, Technology and Industry for National Defence, which reports to the State Council.'[58] China Atomic Energy Authority (CAEA) was created in 1984 to represent China at the International Atomic Energy Agency (IAEA). It used to be an independent agency, reporting to the State Council. Currently, it is integrated into the Ministry of Industry and Information Technology (MIIT). To the outside world, the name of CAEA is still used and represents China at IAEA and other related institutions and treaties, such as the Non-Proliferation Treaty, Nuclear Supply Group and other related multilateral institutions or treaties. Its director is also a deputy minister of MIIT, and heads the downgraded COSTIND, currently a department under MIIT. According to CAEA's website, its main functions include:

- Deliberating and drawing up policies and regulations on peaceful uses of nuclear energy.
- Deliberating and drawing up the development programming, planning and industrial standards for peaceful uses of nuclear energy.
- Organising argumentation and giving approval to China's major nuclear R&D projects; supervising and coordinating the implementation of the major nuclear R&D projects.
- Carrying out nuclear material control, nuclear export supervision and management.
- Dealing with the exchange and cooperation in governments and international organisations, and taking part in IAEA and its activities in the name of the Chinese government.
- Taking the lead to organise the State Committee of Nuclear Accident Coordination, deliberating, drawing up and implementing national plan for nuclear accidents and emergencies.[59]

It also states that its work includes nuclear safety, research and development, the application of nuclear technologies, nuclear energy development in China, and activities with IAEA. In practice, none of these functions or categories of work are taken by or carried out by CAEA, except when representing China at IAEA. The policy-making function is in the hand of NEA; the technical licensing and technical approval are under the National Nuclear Safety Administration, which is not only independent but also under the auspice of the MEP. The international cooperation falls into the jurisdiction of the Ministry of Commerce and sometimes the Ministry of Foreign Affairs is also involved.

Over the past 30 years, the decision-making structure of the nuclear industry has changed significantly, not only because ministries were merged, separated and recreated, but also because the industry was commercialised and corporatized. In 1985, it was the government ministries that were in charge of various segments of the activities of the nuclear industry (see Figure 4.1). In 2010, the state-owned corporations in the industry are accountable to the SASAC while decision-making authorities are shared by various ministries (see Figure 4.2). In both years, efforts were made to create a single authority to coordinate energy policy-making.

Powerful corporations

In contrast to the fragmented and powerless government agencies, the large corporations – the country's powerful elite organisations – have gained substantial economic wealth and political clout. Today they are deeply

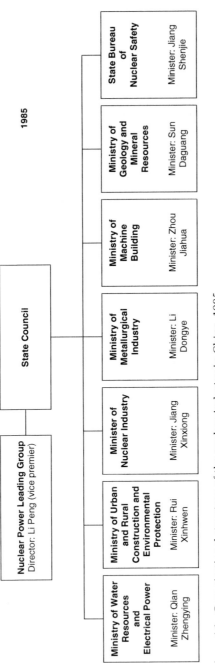

Figure 4.1 Organisational structure of the nuclear industry in China, 1985

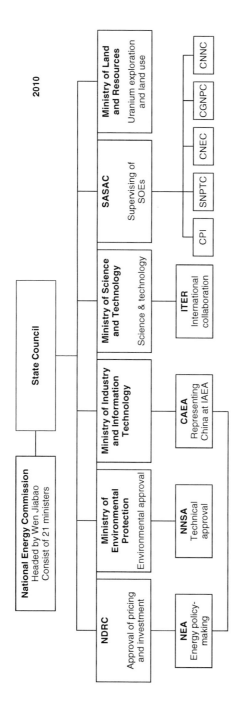

Figure 4.2 Organisational structure of the nuclear industry in China, 2010

enmeshed in a new kind of politics in China. Their monopoly or oligopolistic positions provide them with significant weight in decision making and their economic power has provided them with a good deal of autonomous room to manoeuvre vis-à-vis government agencies. This political clout is not derived from the old planning system; rather it is the by-product of the reform.

In general, enterprises in China have gone through four stages of transformation: 1978–86, 1987–92, 1993–96 and 1997 to the present. In the first stage of reform, while SOEs remained state-owned, their managers had more decision-making autonomy and property ownership was separated from the operation of the company. It was during this stage that private companies and joint ventures first emerged. In the second stage, a contract-based system between the manager and the government was promoted as a new governance mechanism for managing the operation of SOEs.

In 1992, private companies were officially recognised and the government placed a great deal of pressure on SOEs to improve their performance. The third stage started with commercialisation and corporatisation of SOEs and the government's policy of 'grabbing the large, and letting go the small' pushed many small-sized companies into privatising. Large SOEs, especially those in key sectors (including energy), were commercialised and corporatised. Since 1997, concerted efforts have been made by the central government to create 'national champions' in key sectors.

The nuclear industry had been under tremendous pressure to turn around its loss-making situation since the beginning of the reform, but it was not until the end of the 1990s that the State Council decided that the sector would either reinvent itself to join a trial run to become part of the national champions or die by being integrated into other sectors.

In 1999, the government made the fourth significant decision on SOEs. SOEs should focus on core industries, such as electricity, petroleum, telecommunications and banking. All large energy corporations, including the two nuclear corporations, were reorganised and centralised as members of a selected number of China's national champion business groups. Consequently, between 1994 and 2004, the number of SOEs declined from 2 million to under 1 million, and 'by 2006, there were 2856 officially recognised business groups with 27 950 directly owned first tier subsidiaries, employing around 30 million people in China.'[60] Among these groups, 100 are selected as national champions, which include four nuclear companies, CNNC, CGNPC, CNEC and SNPTC, and all major corporations in the electricity sector. In 2009, the central government streamlined the core national champions to strengthen their international competitiveness to 27 so-called 'Chinese backbone corporations' (中国脊梁). CGNPC, but not CNNC, made it onto the list.

The political influence and economic clout allowed their top managers and employees to collect high salaries and rich managerial compensations.

This bred resentment among the general population and sometimes opposition from existing or potential private entrepreneurs, who were blocked by these firms from exploiting lucrative opportunities. Efforts to subject the corporations to increasing transparency and greater regulation have been undermined by the fragmented government agencies and the growing bargaining power of these corporations. Their power rests in their political and economic positions, their control of investment decisions and bargaining power with other sectors and the government. Ironically, they were the products of the government's actions – that is, they were created by the government and were a spin-off from the ministries in the field. Then they were encouraged and supported by the central government to become the national flagship companies.CNNC originated from the Ministry of Nuclear Industry and has evolved into a conglomerate with more than 100 subsidiary companies and a research institute. Until recently it controlled most of the business in the nuclear sector, including research and development (until the Shanghai Nuclear Engineering Research and Design Institute was taken out to be the core of SNPTC), engineering design, uranium mining, and fuel fabrication and fuel cycle services, and nuclear application in medicine and agriculture. The China Institute of Atomic Energy (CIAE) is at the core of its basic research in all aspects of nuclear sciences and engineering. It is responsible for the research and development of faster breed reactors and the back-end of the nuclear fuel cycle. It owns and operates Qinshan I, II and III in Zhejiang and Tianwan in Jiangsu. Its other nuclear power stations under construction include Sanmen in Zhengjian (2 units of AP1000) and Fuqing in Fujian (2 units of CNP1000).

CNEC was split from CNNC in 1998. It is the only corporation in China that is able to build nuclear power stations and indeed has constructed all of them. At the initial state, CNEC went through a similar shedding of employees to the CNNC and almost one-third of its labour force was retrenched between 1999 and 2002. Now it is a monopoly in all aspects of nuclear power station construction. The company has had great experience in the installation of reactors throughout the world – French, Russian and Canadian, as well as Chinese reactors, including the HTR-10. Indeed, it is the only company in China that is licensed to install the components and systems of the nuclear island. Even officials at several international institutions, such as IEA, IAEA and the World Bank, argue that this may be the major obstacle for Chinese nuclear expansion. It has no intention of loosening its grip on this monopoly.

In late 2009, CNEC and CGNPC signed a major contract worth 5.3 billion yuan for CNEC to install eight nuclear reactors (unit 3 and 4 at Ningde, Fujian province, unit 3 and 4 at Yangjiang in Guangdong, and unit 1 and 2 at Fanchenggang in Guangxi autonomous region – all CPR1000 – and 2 units of EPR at Taishan in Guangdong). The core team of CNEC, the China Nuclear Industry 23 Construction Co. Ltd (CNI23), will install all of them.

CNI23 has installed nuclear facilities (regular and nuclear islands and other facilities) of all nuclear power stations. As a broader strategy to promote CPR1000 as the model for China's nuclear expansion, CGNPC needed a team that was able to construct nuclear stations.

In 2004, CGNPC created a subsidiary, China Nuclear Power Engineering Co Ltd. With the permission of SASAC, in 2009, the China Nuclear Power Engineering Co Ltd of CGNPC injected a cash investment into CNI23 so that CGNPC could partake in construction as well. This is an interesting development because, more than a year before this alliance could have developed, the media reported that the central government was considering the possibility of a merger between CNNC and CNEC after a decade of separation. Some board members of CNNC thought the merger would be a natural development because, when CNNC was ordered to split in 1999, two companies were created with their own specialties and no real competition was created. Now, it was argued, they should join forces to better utilise their resources. Some disagree with the alliance, arguing that instead of having fewer players the industry should allow more players to compete so that China can speed up its nuclear energy expansion.[61] As this merger did not take place, SASAC approved CGNPC's investment in CNI23. It has not been made public why SASAC approved the CGNPC's investment in CNEC, but not the merger between CNNC and CNEC.

The fastest-growing nuclear operating company is the China Guangdong Nuclear Power Corporation (CGNPC). It was formally established in 1994 when Daya Bay was connected to the grid and it started its commercial operation with a registered capital of 10 billion yuan. CNNC owns 45% of its shares, Guangdong province owns 45% and CPI has the rest, which used to be in the hands of the Ministry of Electric Power. It is, however, very much a product of the Guangdong government. The nuclear operating company started with nothing except sheer determination. It borrowed to build its first nuclear power station, sold its electricity to Hong Kong, paid back its borrowing and has built huge assets for further expansion. By 2007, it had a total nuclear capacity of 4GWe, almost half of the country's total. It had total asset of 60 billion yuan compared with initial total assets of 3.24 billion yuan in 1994. In 2007, its total profit rose to 3 billion yuan and became the envy of the nation. It was so successful that several banks were chasing it to provide a line of credit for its expansion. CGNPC started issuing corporate bonds. In 2007, the State Council approved an experiment in Guangdong to build a 10 billion yuan fund for future nuclear energy development.[62]

CGNPC now owns and operates Daya Bay and Lingao, and Lingdong (2x1000), Yangjiang (2x1000) and Taishan (2x1700) in Guangdong, Hongyanhe (4x1000) in Liaoning, and Ningde (2x1000) in Fujian is under construction. Construction of its project in Guangxi (Fangchenggang 6x1000) is about to commence.

CGNPC has also expanded to other parts of the country. It signed an agreement with Hubei provincial government to prepare and develop nuclear power stations there, and it also signed a similar agreement with Anhui provincial government. It moved into Jiangsu where Tianwan is already in operation with CNNC as the owner and operator. In addition to nuclear energy, CGNPC has invested in wind, solar, hydro and other renewable energy projects. Finally, it is building a vertically integrated alliance system with its own subsidiaries as well as other major corporations in uranium trading, construction, research and development and personnel training. The simple fact that CGNPC made it onto the list of Chinese Backbone Corporations, but not the CNNC, says a lot about its development and its expansion strategy.

State Nuclear Power Technology Corporation Ltd (SNPTC) was formally created by the State Council in 2007 as one of the elite SOEs in China. The initial investment for SNPTC came from the State Council (2.4 billion yuan, 60%), the large SOEs in the nuclear industry, CNNC, CGNPC, CPI and China National Technical Import and Export Corporation, with 10% each. SNPTC is authorised by the State Council to sign contracts with foreign parties to receive the transferred Generation 3 nuclear power technology, to carry out the relevant engineering design and project management. SNPTC is the key place where Generation 3 technology is introduced, adopted and absorbed. SNPTC is expected to develop a Chinese brand of nuclear reactors through the introduction of foreign technology. Its core is the Shanghai Nuclear Engineering Research and Design Institute – one of the oldest research institutes in China.

Five generating companies have tried to get their foot in the door of nuclear energy development because, in recent years, thermal generation has been a loss-making business for them and it faces a great deal of uncertainty – mainly because of unreliable coal supplies and undependable rail transportation of coal. These generating companies are squeezed in the middle by rising coal prices and they are not allowed to pass on the increases to end-users. Nuclear power stations, so far, are all making money. These power-generating companies are state-owned as well, and like their counterparts in the nuclear sector – CNNC, CNEC and CGNPC – they are business entities, and pursuing profits seems to be their first priority. Nuclear energy development represents new opportunities. The major power generation companies want to compete in the field for the same reasons that the nuclear companies seek to invest in renewable energy. Currently, they all have investment in some nuclear projects, either the ones under construction or those in a state of preparation.

Whether nuclear or power, they are all part of the elite state-owned large corporations in China under the SASAC. These state-owned corporations are in essence amphibious creatures that have never been seen before; the closest relatives are the European trading companies of the 16th–19th centuries, such as Britain's East India Company.'[63] They maintain close ties with the government and often occupy privileged positions in the country's economy. Yet, like the developed world's private giants, they have to compete for their market shares and profits and often behave not much differently from their global counterparts.

They have the political clout and the economic muscle. They are at the vanguard of China's changing corporate culture. These corporations are headed by a group of young, well-educated, urban economic elites – China's 'yuppie corps'. The majority of them have obtained postgraduate degrees, some from overseas. They often have advanced their careers within the same firm or in the same industry. This has allowed many of them to build extensive networks with party and government officials and others in the industry. These top managers have the Party's trust and this is the reason they were appointed in the first place by the CCP personnel department. Many of them are representatives of the National People's Congress and some even serve on the Central Committee of the CCP. However, unlike the old party cadres, they see themselves as managers of modern corporations rather than as ideologues, and their responsibility is to maximise profits of their respective companies, rather than slavishly following the Party's line. They are considered heavyweight players in the Chinese economy.

Second, the top managers of these large state-owned corporations have unique access to information of both governments and markets. Their political appointment by the Party personnel department means that their political status often gives them easy access to information about decision making in both the Party and the government. Their business connections with domestic and international counterparts provides them with access to market information, which is often not obtainable by Party officials. Information is power.

Third, the influence of the large companies rests on their financial power. They are state-owned and have been making profits. Yet, for historical reasons, neither the Ministry of Finance nor SASAC received dividends from them until recently. This practice allowed these large firms to accumulate a large amount of assets. In principle, all nuclear power projects must obtain approval from the State Council. With their ability to make initial investments, either of the three, CNNC, CGNPC or CPI, could align with provinces to start pre-preparation work while lobbying the government for approval. This is an effective way to get their projects approved as it normally takes three to seven years to do pre-preparation work before siting starts. CNNC, CGNPC and CPI can easily align the support from banks, while provincial governments can 'convince' local branches of banks to make financial commitments. Once all the facilities are put in place, it is difficult for the central government to reject the project outright.

Weak and fragmented regulators

Regulation is necessary and important for a market system. Creating regulatory regimes in various sectors is a relatively new phenomenon in China. A regulatory agency in the nuclear sector acts as a representative for the society to ensure that the operation of nuclear facilities does not introduce any undue threats. Nuclear safety builds on a set of general safety principles that are applied in the design and operation of the facilities. 'These principles have been developed over time by analysing failure mechanisms and failure propagation and by applying methods of safety engineering to the technical, human and organisational systems at the nuclear facilities.'[64] When China started its civilian nuclear programme, much of the safety regulation was written by adopting the American codes of safety (see Chapter 6). The country did create a regulatory agency in 1984 to ensure safety of nuclear power plants as early as 1984. The National Nuclear Safety Administration (NNSA) was created as an independent agency to issue licences and grant technical approvals. The challenge China currently faces is the expansion of a team of professional regulators.

In addition to safety regulations, economic regulation has to be in place – that is regulation of prices and other commercial terms, investment and service quality. Regulation is necessary because: (a) the behaviour of 'natural monopolies' needs to be controlled to prevent monopoly enterprises from raising prices at will while reducing the electricity supply services; (b) fair and efficient competition is needed; and (c) public interests need to be promoted and protected. It is a great challenge to have an effective regulatory regime while encouraging competition in the electricity sector. In China, the challenge is to distinguish between regulatory and decision-making functions and to encourage those traditional decision-making institutions to transfer their regulatory functions to the regulators. This has proved extremely difficult.

The Chinese government has invited multilateral organisations, such as the World Bank and Asian Development Bank, foreign regulatory agencies and international consulting firms to help draft regulations and create regulatory regimes. The effort has not been as successful as the efforts in creating safety regulations in the nuclear industry. This is primarily because the institutions holding decision-making authorities, such as NDRC or the Ministry of Finance, were reluctant to give up their regulatory functions to the newly created regulatory agencies. Consequently, there developed three new problems:

First, policymaking and regulatory functions are mixed.

Second, planning approvals and investment supervision functions are confused.

Third, tariff regulation functions are not allocated effectively.[65]

A study conducted by the IEA and a joint study of the World Bank and international consultants show that, despite the success in reforming the electricity sector in China, there is still little distinction between direct control and regulation, and the existing distribution of regulatory functions is inefficient and inconsistent. When market forces are introduced but a regulatory system is not in place, government agencies have constantly tried to rein in SOEs with little success or effect. For example, when competition among the companies led to excessive investment in power generation, the multiplication and competition among different government agencies undermined the state's capacity to manage this crucial sector.

In 2002, four main agencies were mandated to oversee the development of the power sector: the NDRC; the newly created State Electricity Regulatory Commission (SERC); the State Asset Supervision and Administration Commission (SASAC) and the State Environmental Protection Administration (SEPA). The NDRC nonetheless is not a regulator. It is China's main macroeconomic planning and monitoring agency under the State Council. After 2002, it insisted it maintain its authority over investment approvals in the power sector and over electricity pricing. This has created difficulties in pricing and investment because in a market system pricing can no longer be an effective instrument to ensure macroeconomic stability as in a planned economy. Therefore, electricity tariff regulation should not be confused with economic macro-regulation. 'Effective power price regulation may be possible only when the regulator is in charge, so that decisions are not influenced by political considerations.'[66]

- SERC was established in 2002 as an independent regulatory agency for the newly liberalised power sector, but it was not granted the powers that were crucial for the job. SERC is responsible for establishing and overseeing market rules, including competitive bidding rules and those protecting fair competition. It is in charge of supervising market entry and licensing, yet it does not have control over project or investment approval. It is asked to propose modifications to power tariffs but does not set tariffs. It is also supposed to participate in and regulate technical standard setting, yet it has neither the manpower nor expertise to do so. In sum, without any authority over regulating investment and pricing, SERC, the sector's independent regulator, is an empty shell.
- SASAC was created in 2003 to 'own' and manage state-owned corporations in the public interest. As the ultimate owner of state-owned corporations, SASAC is to oversee the largest and centrally owned non-financial SOEs, a total of 189 entities initially. SASAC is under the direct authority of the State Council, and its main responsibilities are defined as (a) to carry out its responsibilities as investor and to guide and promote the reform and reorganisation of the SOEs; (b) to represent the state on the supervisory boards of some large enterprises; (c) to appoint, dismiss and assess senior

executives and to assign rewards and penalties on the basis of performance; (d) to monitor the extent to which SOE value is maintained or enhanced; (e) to draft laws and regulations on the administration of SOEs and set related rules and regulations; (f) to direct and supervise the administration of SOEs under local ownership in conformity with the law.

The fundamental idea underpinning the SASAC is to exercise ownership rights in a centralised and unified manner, and according to Company Law it has not developed into a full-fledged regulatory agency yet. For example, it seeks to ensure high returns on state assets, but it does not have the authority to collect dividends from SOEs. That power is in the hands of the Ministry of Finance. Neither does it have direct control over their budgets. Consequently, SASAC has few resources and little power to ensure the value of state assets or to monitor these state-owned corporations. SASAC is mandated to supervise the business performance of these large corporations, but it does not have control over the appointment of senior people in these companies. This power vests with the Organisation Bureau of the CCP Central Committee. SASAC only makes the less important personnel decisions, while the Party appoints the senior managers. However, unlike SASAC, the Organisation Bureau of the CCP Central Committee has no authority over the actual operation of these corporations

- SEPA – in March 2008, the State Environment Protection Administration was upgraded into a full-ministerial agency, the Ministry of Environmental Protection (MEP). This is a clear indication of how environmental problems have caused serious concerns for the central government. For the first time, the National Nuclear Safety Administration was integrated into the SEPA. The chairman of NNSA is the deputy minister of SEPA. NNSA was created in 1984 to ensure the safety of nuclear power plants in terms of construction and operation. It was accountable to the State Council until 2008, when it was placed under the MEP. All nuclear power plants need to obtain approval from both NNSA and MEP: NNSA issues the licence and approves the project based on technical and safety standards, while MEP approves the project based on environmental assessment.

MEP should be the key player in ensuring the protection of the environment, especially in the electricity sector, which contributes 40–60% of the country's GHG emissions. While NDRC considers the approval of a project, the power company is required to seek approval from SEPA simultaneously. The environmental assessment and approval are integral parts of all nuclear power projects. One of the main problems the MEP has is the lack of human or financial capacity to sanction erring companies and a lack of the authority to repudiate the project approvals granted by the NDRC. MEP has a limited manpower of 2600, with only 300 working out of the headquarters in Beijing. The US EPA has 17 000 employees, nearly 9000 of whom work in Washington, DC.

In dealing with large energy corporations and local governments, the MEP seems to be dwarfed by both. Making profits and maximising local development are the first priority, and both the energy corporations and local governments have much more resources at hand in dealing with the MEP. Finally, the MEP is not as heavyweight a player in economic decision making as, for example, is NDRC or the Ministry of Finance. So frustrated by the spread of environmental pollution, the MEP minister was recorded criticising 'the lack of respect for the environment as the country carries out its economic stimulus plan,' and some government officials, 'what is the point of driving a Mercedes Benz while breathing dirty air and drinking dirty water.'

Provincial and local governments have powerful political interests in the power sector that often work against effective regulation. Even though the main power companies are owned by the central government, they all run large numbers of subsidiaries across the country. These are often subject to political pressure by the host provincial governments while being lured by the financial incentives given by them to invest and expand their local operations, even, at times, in violation of the 'rules' set by the central government. Every provincial government has an interest in making sure that it has sufficient electricity supply and wants the benefits from job opportunities offered by the industry. Holding exactly the same bureaucratic rank as provincial governments, central ministries are explicitly forbidden to issue instructions to their local counterparts. The Politburo and the State Council are the ultimate authority for battles between central regulators and provincial governments. Yet, neither have the human resources nor the ability to deal with every power project in question.

Conclusion

China 'has opened up to and is being buffered by far more pluralisation and omnidirectional influences.'[67] This has also been the case with nuclear energy decision making. The old bargaining dominated by three pillars – the Chinese Communist Party (CCP), the military and the scientific community – was considerably broadened to include not only relevant government agencies, powerful corporations and provinces, but also think tanks and organised or semi-organised groups. Facing the two challenges – energy security and climate change – the central government sees nuclear energy as an alternative, and everyone wants to have a say on what strategy the country should adopt regarding nuclear energy development, at what speed nuclear energy should develop, where power plants should be located, or whose technology the country should adopt in its nuclear energy expansion. Nuclear energy has come a long way from a sector no one wanted to host, to an integrated part of the energy industry.

Given that the Chinese energy institutional landscape is characterised by overlapping jurisdictions and inconsistent waves of centralisation and decentralisation, 'energy policy in China today is a battleground of negotiation among power actors with conflicting interests that are in evidence at all levels of analysis.'[68] The failure of the National Energy Leading Group, formed in 2005, to coordinate energy policies led to the creation of the National Energy Administration in 2008, under which nuclear was brought in. Yet with growing political and economic influence, some have argued, China's large state-owned energy corporations have been driving the national energy priorities and policies. Given these energy corporations have their own distinct interests to promote in the decision-making process, it is not difficult to see the complexity of the bargaining process.

Notes

1. Strauss and Saavedra, 2009, p.552.
2. Mertha, 2009, p.995.
3. Qi Zhou, 'Organisation, Structure and Image in the Making of Chinese Foreign Policy since the Early 1990s', PhD thesis, Johns Hopkins University, 2008, p.131.
4. Dumbaugh and Martin, 2009, p.20.
5. Edward A. Cunningham, 'China's Energy Policymaking Structure and Reforms', Testimony before the US-China Economic and Security Review Commission, 13 August 2008.
6. Lieberthal and Oksenberg, 1988, p.15.
7. Fewsmith, 2008.
8. Downs, 2006.
9. Hecht, 1998, p.4.
10. Pierson, 1993, p.596.
11. Feng Fei, 'A Study of Financial, Taxation, and Economic Policies for Sustainable Energy Development', at the Eighth Senior Policy Advisory Council Meeting, The Great Hall of the People, Beijing, China, 18 November 2005, p.39.
12. NEA, 2004, p.13.
13. Winskel, 2002; Hecht, 1998.
14. Cheung, 2009, p.52.
15. Frankenstein, 1999, p.208.
16. Lieberthal and Oksenberg, 1988, p.38.
17. 李鹏, 2004, p.66.
18. Naughton, 1988, pp.351–86.
19. Feigenbaum, 2003; Lewis and Xue, 1988.
20. 孟戈非, 2002.
21. 当代中国的核工业, 1987, p.80.
22. 'Foreign Funding Sought for Peaceful Use of Nuclear Industry', Xinhua News Agency, 3 September 1990; 'Foreign Exchange and Cooperation by China's Nuclear Industry; *Liaowang* (Overseas edition, Hong Kong), 11, 16 March 1987, pp.35–36; Michael G. Gallagher, 'Nuclear Power and Mainland China's Energy Future', *Issues and Studies*, 26:12 (1990), pp.100–120; 'Storing of Nuclear Wastes on Chinese Mainland', *Issues & Studies*, 20:3, 1984, pp.8–10.

23. Richard A. Bitzinger, 'Dual-Use Technologies, Civil-Military Integration, and China's Defence Industry', in Nan Li (ed.) *Chinese Civil-Military Relations: the transformation of the People's Liberation Army*. London: Routledge, 2006, p.181.
24. Huang, 1996; Yang, 2004.
25. Frankenstein and Gill, 1996, p.419.
26. Quoted from 'Energy: Converting the Nuclear Industry to Civilian Use', BBC Monitoring Service: Asia-Pacific, 13 February 1985.
27. Cheung, 2009; 李鹏, 2004.
28. Richard Baum, 'China in 1985: The Greening of the Revolution', *Asian Survey*, 26:1, 1986, p.30.
29. Shambaugh, 1996.
30. Quoted from Esposito, 1991, p.393.
31. Ibid, p.386.
32. Chow, 1997, p.406.
33. Andrew-Speed, 2004.
34. Downs, 2000; Downs, 2004:25; Downs, 2006.
35. Lu, 1993, pp.5–71.
36. Zhang 1997, p.204.
37. Hu, 1994, p.4. When COSTIND was created in 1982, it might have had some political clout as it was in charge of coordinating all defence industries. 'China's economic reforms, changes in the leadership, and continuing systemic problems in the defence sector wakened COSTIND's influence' (Shirley A. Kan, 'China: Commission of Science, Technology, and Industry for National Defence (COSTIND) and Defence Industries', CSR Report to Congress, 3 December 1997, p.12). Its influence was further undermined in the late 1990s when 'Chinese leaders transformed the COSTIND into a civilian entity to bring together the administrative and regulatory functions... [while] most of the other functions of the old COSTIND went to the General Armaments Development of the PLA' (Richard T. Cupitt, 'Non-proliferation Export Controls in the People's Republic of China' in M.D. Beck, R.T. Cupitt, S. Gahlaut, S.A. Jones (eds) *To Supply or to Deny: Comparing Non-Proliferation Export Controls in Five Key Countries*, The Hague: Kluwer Law International, 2003, p.127).
38. Readon-Anderson, 1987, p.41.
39. Chow, 1997, p.406.
40. '为深化改革建言，为核工业发展献策', 中国核工业 ('The Information and development of the Chinese Nuclear Industry: Opinions and Advice', *China Nuclear Industry*, 2, 1998, pp.4–15.
41. 郑庆云, '浅谈 "硬道理"与 "新思路"', 核经济研究, Zheng Qingyun, 'The Hard Truth and a New Thinking', *Nuclear Economics Research*, 4:3, 1994, p.7.
42. Mark Hibbs and Ann MacLachlan, 'Citing Cost, Qinshan-1 Repair, Beijing Balks at Liability Regime', *Nucleonics Weekly*, 40:20, 20 May 1999, p.1.
43. Noureddine Berrah, Ranjit Lamech and Jianping Zhao, *Fostering Competition in China's Power Markets*, Washington, DC: The World Bank, March 2001, p.7.
44. Lieberthal and Oksenberg, 1988, p.51.
45. Li made three suggestions for electricity reform in 1983: (a) 'walking on two legs' – planned and market ways of development, (b) two management systems – central government in charge of grid and large power project developments while provincial and local governments investing in smaller generation projects and (c) generating profits for its own development (以电养电), a concept that was eventually adopted by the nuclear development in Guangdong. See 李鹏, 2004:125 and Xu, 2002.

46. Lieberthal and Oksenberg, 1988, p.56.

47. Lu, 1993, p.57.

48. Smil, 1981; Planning Division of the Ministry of Electric Power, 'Report on Electricity Supply and Demand in China' in Planning Division of the Ministry of Electric Power (ed.) *Reform and Planning in the Power Industry*, Beijing: China Electric Power Press, 1996, pp.686–88; Chu-yuan Cheng, *The Demand and Supply of Primary Energy in Mainland China*, Taipei, Taiwan: Chung-Hua Institutions for Economic Research, 1984, p.87; Anonymous, 'Priority to Developing Energy', *Beijing Review*, 7 November 1983, p.23.

49. Lu, 1993; Xu, 2002; 李鹏, 2004.

50. Xu, 2002, 2004.

51. Lieberthal and Oksenberg, 1988, p.22.

52. Price, 1990, p.48.

53. Quoted from Qi Zhou, 'Organisation, Structure and Image in the Making of Chinese Foreign Policy since the Early 1990s', PhD dissertation submitted to Johns Hopkins University, 2008, p.138.

54. Matthew J. Matthews, 'Nuclear Power Shapes Up', *China Business Review*, 12, July–August 1985, pp. 23–27.

55. Quoted from Qi Zhou, 'Organisation, Structure and Image in the Making of Chinese Foreign Policy since the Early 1990s', PhD dissertation submitted to Johns Hopkins University, 2008, p.141.

56. Erica Downs, 'China's "New" Energy Administration', *China Business Review*, 35:6, 2008, p.42.

57. Ibid.

58. Andrew C. Kadak, 'China's Energy Policies and Their Environmental Impacts'. Testimony before the US-China Economic and Security Commission, 13 April 2008.

59. http://www.caea.gov.cn/n602670/n621894/n621895/32165.html

60. Paul Guest and Dylan Sutherland, 'The Impact of Business Group Affiliation on Performance: Evidence from China's "National Champions"', *Cambridge Journal of Economics*, advance access published 8 May 2009.

61. 明茜, 竞争还是整合? 中核, 中核建设合并尚无定论, 21世纪经济报道, (Ming Qian, 'Competition or Merger: CNNC and CGNPC', *21st Century Economic Report*, 28 August 2008, p.18.

62. 冯志卿, 中广核'以核养核' 中国投资, (Feng Zhiqing, 'CGNPC's Rolling Development', *China Investment*, 3, 2008, pp.66–70.

63. 'The World Turned Upside Down', *The Economist*, 17 April 2010, p.11.

64. Björn Wahlström, 'Reflections on Regulatory Oversight of Nuclear Power Plants', *International Journal of Nuclear Law*, 1:4, 2007, p.346.

65. SERC, Ministry of Finance and World Bank, 'Study of Capacity Building of the Electricity Regulatory Agency, P.R. China', The World Bank, 21 August 2007, p.121.

66. IEA, 2006b, p.81.

67. Gill, 2001, p.258.

68. Edward A. Cunninghan, 'China's Energy Policymaking Structure and Reforms', testimony before the US-China Economic Security Commission on 'China's Energy Policies and Environmental Impacts', 13 August 2008.

5
Who Pays? The Economics of Nuclear Energy

The cost of nuclear power has been identified as one of the determining factors for the future of nuclear energy development and expansion. Some argue that it does not make economic sense to build NPPs, given that the required capital investment is much higher than for other forms of electricity generation, that it takes much longer to build NPPs, and their financing is subject to far more uncertainties. Others argue that the intensive capital investment can be compensated by low fuel and operating costs. New NPPs can be competitive when fossil-fuel prices rise and potential external costs are taken into consideration. These external costs could include air pollution, greenhouse gas emissions, import dependence, the uncertainties of fuel costs and the comparative risks of different alternatives. To some, 'it is clear that nuclear power has been unusually difficult to assess in economic terms'[1] because national or even local conditions vary, including the costs of capital, labour and materials, the regulatory environment, and the availability and costs of alternative generating technologies. More importantly, political acceptance of, and public opinion on, nuclear energy can throw any economic calculation off balance.

Nuclear energy development takes place in a macroeconomic context and hence the issue is not only what should be included in the cost calculation, but also how it can be financed (by private or public, domestic or foreign, equity or debt) and under what conditions. Many developing and transition economies desiring to develop or expand nuclear energy, including China, face another major challenge – competing demands for investment. Since multilateral and regional financial institutions, such as the World Bank and the ADB, are prohibited from making concession loans to nuclear energy projects, putting resources, especially foreign exchanges, into nuclear projects is competing with the pressing needs for health, education, poverty alleviation and environmental protection.

The questions of where to put the resources and how resources are allocated are political issues that require more than a simple economic calculation. In an increasingly competitive market system, investors (private as well as public) are unwilling and unable to invest in NPPs unless they can ensure at least the cost can be recovered and with a reasonable rate of returns. Therefore, electricity prices must be set high enough to allow the investors to achieve these objectives.

This chapter explores the economic challenges in nuclear energy expansion from two angles: the investment and operational costs of a NPP, and power tariffs setting. Price setting in conjunction with investment planning is the most powerful instrument available to governments in managing the economy. They have been thorny political and social questions. Once the economic reform started in the late 1970s and early 1980s, government budgetary financing was gradually replaced by credit allocation through the banking system. By acceding to a dramatic reduction in its direct control over investment resources, the central government lowered entry barriers step by step across sectors. Electricity generation was open to other sources of investment at the early stage of the reform. Pricing was another instrument the government used to shape economic incentives and to attract investment in the sector. Price reform was not as drastic as that in investment, especially not in the electricity sector. Power pricing remains under tight government regulation. Yet, as Douglass C. North points out, 'a change in relative price leads one or both parties to an exchange, whether it is political or economic, to perceive that either or both could do better with an altered arrangement or contract.'[2] Renegotiating a new arrangement leads to a change in formal rules and laws and also a gradual erosion or replacement of the old norms. Thus, price changes alter the incentive structure as well as the organisational structure of players.

Economic rules are made in the political arena and changes in pricing and investment have mirrored domestic and international political changes. An analysis of the political processes involved in making these decisions is at the core of this chapter. The government, through its central planning agency (SPC and then NDRC), its regulatory agency (the State Electricity Regulatory Commission, SERC) and other concerned agencies has been trying to reorganise the pricing system for electricity.

The 1992 Price Law was revised in 1998 but the section concerning electricity was not changed. Power tariffs were still under the category of government-set prices, determined by the relevant government departments at the national and provincial levels. In 2003, the State Council, in its document 'Scheme for Power Price Reform', tabled proposals for further reform linked to the development of competitive regional power markets. A follow-up document in March 2005 set out the plans in more details, but only moderate changes were made, especially on end-user pricing. Pricing for nuclear power is part of the general development in electricity tariff reform that can only be achieved in conjunction with economic and political reform in the industry and in the economy in general.

How to calculate the cost?

The capital investment required to construct a NPP typically represents some 60% of the total cost of nuclear electricity generation and this high cost of capital can be offset by low and more stable costs of fuel (15%) and of operation and management (O&M) (25%). Most economic arguments for or against nuclear power tend to focus on the factors that may affect capital investment, such as the length of construction that is 'a source of extra cost due to the interim interest charges on investment capital,'[3] and exchange rates for developing and transition economies, the rising cost of materials for construction, such as steel or non-iron materials, which rose by 50–100% in 2007 alone, depending on location, quality and quantity.[4]

The capital-intensive nature of the nuclear industry, however, is not unique in the electricity industry; major hydro projects carry costs per kilowatt installed capacity that are of the same order of magnitude. It is also a fundamental feature in other segments of the electricity industry, such as transmission and distribution networks. The initial investment is high and long term (a large hydro project often takes much longer to build than a NPP); it is specific, and therefore asset-sunk (once investment is made in any of the segments of the power industry, it cannot easily be refitted for another operating environment); and it is highly interdependent – investors would not invest in generation capacity unless they have guaranteed access to the grid and to future customers. For these reasons, power plants, including nuclear ones, used to be financed by governments directly or with government-guaranteed financing.

When China started its first NPP (Qinshan I), neither the cost nor the investment was an issue because it was approved as a political project. Nonetheless, debates on the economic merits of nuclear power projects were soon underway. Since the beginning, the argument that nuclear energy would not be economically competitive in China has been upheld by many policy makers, people in the electricity sector and scholars of macroeconomics because: (a) China has abundant coal reserves; and (b) as a developing country, China faces multiple competing demands for resources. This may explain, at least partially, the reluctance of the central government to put resources into nuclear energy development and the unwillingness of the power industry to invest in nuclear power plants until the 21st century.

While those against nuclear energy tend to focus their argument on the initial intensive investment and the costs of managing nuclear power plants' after-life, those in the nuclear community have repeatedly emphasised the low cost over the lifetime of nuclear power plants. Indeed, the latter were quite optimistic about nuclear energy development, both in terms of technology and economics. 'In 1980, after exhaustive investigations and analyses were carried out by the Nuclear Energy Investigation and Research Group of the State Science and Technology Commission, it was concluded that while we had the technology to construct nuclear power plants, the difference in investments, together with the investments in the systems needed for the maintenance of the operation between the large nuclear power plants in the southeast and the coal-fired power plants of the same capacity is negligible.'[5]

Two years later, a joint working group – the China Electrical Engineering Society, the Chinese Nuclear Society and the China Mechanical Engineering Society – published an additional estimate of the costs of NPPs. A number of articles appeared in several Chinese newspapers and magazines comparing the cost of electricity from coal-fired and oil-fired thermal power plants with that of nuclear power. The general conclusion was that the cost for a kilowatt of electricity from nuclear power stations could be '18% cheaper than the power station fired by coal and 59% cheaper than the power station fired by oil.'[6]

Studies were also published on comparative costs of nuclear power plants based on their size. Table 5.1 also shows that the larger the unit, the more efficient it was.

This calculation was made based on the domestically designed pressurised water reactors with 300MW and 125MW capacity. The investment cost (including fuel) would be 2500 yuan and 2130 yuan per kilowatt, respectively. The estimated cost for a unit of 1000MW was lower, at 1500 yuan per kilowatt, 'based on the condition that they would be in serial production after the technology to construct them has been acquired.'[7]

In 1982, a MEP's study, 'Provisional Regulations for the Economic Analysis of Electric Power Engineering', concurred that it was more economical to construct nuclear power stations with 1000MW capacity than coal-fired power plants of the same capacity in Shanghai. The conclusion was drawn from the fact that both coal and transport prices were at the time too low and would soon see a significant increase.

The optimistic calculation of nuclear power was not a dream only of Chinese economists and nuclear scientists. In the 1970s, many countries in the West, where nuclear power was considered as a solution to the energy crisis during the oil crises, were of the same opinion, especially when NPPs could be installed with standardised technology and equipment.

Table 5.1   Costs of nuclear energy calculated in 1982

Capacity of a unit	1000MW	300MW	125MW
Investment yuan/kw (including fuel)	1500	2500	2130
Capacity factor (%)	70	70	70
% of electricity consumption by the plant	5	7	7

Source: Luo Anren, 'Economic Advantages of Nuclear vs. Coal-Fired Power Plants Weighted', *Nuclear Power Engineering*, 4: 6, 1983, p.144.

Nonetheless, this optimistic argument on nuclear energy was drowned out by those in favour of coal. In September 1984, Yang Haiqun of the SPC's Economic Institute, published an article, 'On the Decline of the World Nuclear Energy', arguing that China should abandon its nuclear development plan in the interest of both economics and safety.[8] In his article, Yang pointed to the economic challenges the nuclear industry was facing: intensive initial capital investment, long construction periods, competition for resources among sectors, low load factor, and high and uncertain costs for decommissioning and waste disposal. According to Yang, the construction costs for NPPs in 1983 were at least 60% higher than those for coal-fired capacity in developed countries, and the costs would be even higher in China where coal supply was abundant. That this article was published at all at the beginning of the reform, when different ideas challenging the policies were rarely made openly, highlights the internal debates among top policy makers. Yang Haiqun's argument was used repeatedly by those who had doubts on nuclear energy.[9]

By the early 1990s, as the central government insisted on making 'electric power the core' in its energy policy, many argued that nuclear power development made even less economic sense than in the early 1980s, for several reasons.[10] First, the economic challenges NPPs faced earlier remained: high capital investment, and long construction periods compounded with high interest rates. Second, competition for resources became an even more serious challenge as generation capacities had expanded but transmission and distribution had fallen behind. Without adequate investment in transmission and distribution, investing in nuclear power generation made no economic sense. Third, throughout the 1980s and 1990s, China had been going through an expansion–contraction cycle and this made investment decisions on nuclear energy very difficult. Indeed, decisions on all large investments were subject to rapid macroeconomic and political changes. Finally, until the early 2000s, coal prices had remained low in China and the world. As IEA suggested in 2006, so long as coal prices remained below US$70/tonne, nuclear power would not be economically competitive.[11]

A number of changes in the past few years have led to a growing interest in nuclear energy. One was the rising coal prices and tight coal supplies. The average annual increase in coal prices between 2000 and 2006 was 16.84%, compared with consumer price index (CPI) being a little over 1% of the same period (see Table 5.2).

During the same period, tight domestic coal supplies meant more coal imports (see Table 5.3) and higher coal prices in the world market. The price of thermal coal from Australia rose steadily from US$51/tonne in 2005 to US$70/tonne in 2007, US$136/tonne in 2008 and back down toUS$77 in 2009 in the midst of the global financial crisis. As soon as the economy recovered, coal price at the world market jumped to US$95/tonne in the first week of 2010 and reached $100/tonne 10 days later.

Table 5.2 Annual change in consumer price and coal price (previous year = 100)

	2000	2001	2002	2003	2004	2005	2006
CPI	100.4	100.7	99.2	101.2	103.9	101.8	101.5
Coal price	119.8	109.7	111.6	109.7	130.2	128.7	108.2

Source: National Bureau of Statistics of China, *China Statistical Yearbook 2008*, Beijing: China Statistics Press, 2009; 崔民选, 中国能源发展报告, 2008, 北京: 社会科学文献出版社, 2008, p.55 (Blue Book of Energy: Annual Report of China's Energy Development, 2008).

Table 5.3 Coal imports, 2000–07 (million tonnes)

2000	2001	2002	2003	2004	2005	2006	2007
2.18	2.49	10.81	11.03	18.61	26.17	38.25	52.00

Source: 崔民选, 中国能源发展报告, 2008, 北京: 社会科学文献出版社, 2008, p.59 (Blue Book of Energy: Annual Report of China's Energy Development, 2008).

The rising coal imports took place at a time when study after study showed that cheap, better quality and more accessible coal is disappearing quickly in China. Coal depletion presents an 'energy security' issue to both central and provincial policy makers. It also means the trend of coal prices is moving upwards rather than downwards. Economic incentive structures are increasingly shifting in favour of nuclear energy.

A second important consideration relates to climate change. Climate change is real and it is now a widely held view that energy production and consumption contribute the lion's share of changes to our planet's climate. This means: (a) low-carbon base-load power must be developed; and (b) the costs of environmental pollution incurred in energy production and consumption must be calculated in energy prices. 'The economics of nuclear relative to fossil-fuelled generation, particularly coal, improves with carbon pricing.'[12]

Several studies on comparative costs of nuclear and thermal power carried out by various foreign institutions were introduced to support the arguments for and against nuclear development in China. Studies done by MIT in 2003 and 2009 were taken as authoritative references in supporting NPP development in China (see Table 5.4). The studies show that the costs of generating a kwh of electricity for base-load nuclear, coal and natural gas generating technologies had gone up during the period under study, and so had the capital cost for the three technologies, expressed as an overnight cost per unit of capacity.

The overnight cost for the construction of a new nuclear power plant doubled from US$2000/kw in 2003 to US$4000/kw in 2007. The overnight cost for coal-fired and gas-fired thermal plants also increased, but not as much as the increase of nuclear power. The studies also showed a substantial increase in carbon emission costs.

Table 5.4   Costs of electric generation alternatives

2003 study (in US$ 2002)	Overnight cost[a] (US$/kW)	Fuel cost (US$/ mmBtu[b])	Base cost	Levelised cost of electricity (¢/kwh)	
				With carbon charge US$25/tCO$_2$	With same cost of capital
Nuclear	2000	0.47	6.7		5.5
Coal	1300	1.20	4.3	6.4	
Gas	500	3.50	4.1	5.1	
2009 update (in US$ 2007)					
Nuclear	4000	0.67	8.4		6.6
Coal	2300	2.60	6.2	8.3	
Gas	850	7.00	6.5	7.4	

Note: [a] The overnight construction cost, defined as the total of all costs for constructing a power plant as if they were spent instantaneously, tends to be high for nuclear power plants.
[b] MMBtu – million metric British thermal units.

Source: MIT, 'Update of the MIT 2003: Future of Nuclear Power', 2009, p.6.

Table 5.5   Costs of electric generation alternatives, including carbon charge

In US$ 2007	Overnight cost (US$/kw)	Fuel cost (US$/ mmBtu)	Levelised cost of electricity (c/kwh)	
			With carbon charge US$25/tCO$_2$	With carbon charge US$50/tCO$_2$
Nuclear	4000	0.67	8.4	8.4
Coal (low)	2300	1.60	7.3	9.4
Coal (moderate)	2300	2.60	8.3	10.4
Coal (high)	2300	3.60	9.3	11.4
Gas (low)	850	4.00	5.1	6.0
Gas (moderate)	850	7.00	7.4	8.3
Gas (high)	850	10.00	9.6	10.5

Source: Paul Joskow and John E. Parsons, 'The Economic Future of Nuclear Power', *Daedalus*, 138:4, 2009, p.54.

In 2009, Professor Paul Joskow from MIT published another more updated comparative study on costs, taking into consideration the different prices for coal and natural gas and the price for carbon emission, which will change the whole scenario for nuclear energy development.

Table 5.5 shows that the cost of nuclear power would be competitive against the cost of power from coal at a moderate price even at the lower charge of US$25/metric tonne of CO_2. With $50/metric tonne of CO_2, nuclear power would be cheaper than all options except natural gas. 'These numbers illustrate the tradeoffs facing an investor making a choice on which type of capacity to install.'[13]

Other studies have also been used in the analyses and debates in China. The French Energy Secretary released a study in 2003, stating that capital cost for EPR was about €1650–1700/kw, compared with €1200–1400/kw for a coal plant. The capital cost of CANDU 6 in Ontario was Can $2972/kw, compared with Can $1600 for coal, which is abundant in Ontario.[14] The costs of GHG emissions have become an integral part of the cost calculation. 'The August 2003 figure put nuclear costs at €2.37c/kwh, coal €2.81c/kwh and natural gas at €3.23c/kwh (on the basis of 91% capacity factor, 5% interest rates, 40 year plant life),' according to the World Nuclear Association. If the price of CO_2 is €20, the electricity price of coal generation would reach €4.53c/kwh,[15] far higher than nuclear power. This made nuclear energy very competitive with other electricity generation technologies.

This conclusion was confirmed by the US Congressional Budget Office, which in 2008 provided its own reference scenario for comparing the costs of electricity from nuclear, coal and gas power plants (see Table 5.6).

For NPPs, any cost figures normally include spent fuel management, plant decommissioning and final waste proposal. 'These costs, while usually external for other technologies, are internal for nuclear power (i.e., they have to be paid or set aside securely by the utility generating the power, and the cost passed on to the customer in the actual tariff).'[16] These 'external' costs vary considerably (about 9–15% of the initial capital cost of a NPP) and have already been accepted as part of the integrated costs of nuclear power

Table 5.6   Key assumptions underlying CBO's reference scenario

	Advanced nuclear	Conventional coal	Conventional gas	Innovative coal	Innovative natural gas
Construction					
Time (year)	6	4	3	4	3
Overnight costs (US$1000/MW)[a]	2358	1499	685	2471	1388
Operating costs					
Fuel (US$/MWh)[b]	8	16	40	17	52
Fixed O&M (US$/MWh)	8	4	1	6	3

Note: [a] Overnight construction costs do not include financing charges.
[b] Fuel costs include a US$1 per megawatt hour charge to cover the cost of spent fuel disposal.

Source: Congressional Budget Office, 'Nuclear Power's Role in Generating Electricity', May 2008, p.19.

generation. Recent studies, as indicated above, all include costs of GHG emissions. The message is clear: if carbon dioxide charges are to be imposed in the future, these costs would have to be included in the calculation and the utilities would have to assume the full cost of their fuel choice – that is, they would have to pay for the damage inflicted by emitting carbon dioxide. Once the costs of carbon emissions are incorporated, nuclear energy would become economically competitive. This would have significant effects on investment decisions and plant dispatch.

Meanwhile, 'the cost of nuclear power is highly sensitive to its capital costs (both the absolute levels of capital expenditure and the cost of capital) because of high capital intensity (typically that cost of a new 1600MW plant is likely to exceed $5 billion) and long lead times.'[17] Interest rates, availability of capital, charges on carbon emissions, and construction time are among the major hurdles for nuclear development. Furthermore, costs vary greatly because 'labour and material costs vary and their impact varies with the localisation rate (i.e., the percentage of plant components that are locally manufactured or procured).[18] Uncertainty is the most serious challenge to investors.

In China, costs have also gone up significantly because of rising labour and land costs in the past two decades (see Table 5.7). While they remain below the costs in developed countries, the capital costs for NPPs currently in operation vary considerably, depending on the technology used and especially the proportion of the indigenous technology, materials and workforce used to build them. For example, both Lingao and Qinshan III nuclear used the imported technology – French and Canadian – and their overnight costs were similar, about RMB12 000/kw, compared with the overnight cost of RMB5000/kw for coal-fired capacity.

Costs for the first-of-a-kind nuclear project are always much higher, sometimes 50% higher, than the following projects. Daya Bay was the first commercial NPP in China with the imported turnkey technology, and it was constructed by companies from France, UK and elsewhere. Consequently,

Table 5.7   Costs of nuclear power plants in China

NPP	Capacity	Total investment (US$ million)	Overnight cost (US$/KW)
Daya Bay	2x984MW	4400	2236
Lingao I	2x900MW	3640	1838
Tianwan	2x1060MW	3200	1509
Qinshan I	1x300MW	210	685
Qinshan II	2x642MW	1779	1386
Qinshan III	2x728MW	2800	1923

Source: 陈村兰, '基于工程造价及发电成本的核电于火电比较研究', 科技与管理, (Chen Chenlan, 'Comparing the Investment and Overnight Cost of Nuclear and Coal Energy', *Science, Technology and Management*, 32:4, 2005, p.6.

its overnight cost was much higher than the world average at the time. The Nuclear Energy Agency of OECD (NEA) pointed out that using standardised plant designs and constructing multiple units on a single site can reduce the capital investment costs.[19] An earlier study showed that capital cost savings obtained by standardisation and series construction could range from 15% to 40%.[20] Lingao was a duplicate of the Daya Bay NPPs. With 30% of domestic contribution in equipment and labour, the cost was nearly 20% lower than that of Daya Bay. Qinshan I and II used indigenous technology and construction even though both had imported components and the cost was significantly lower than if they had used imported technology and equipment.

Qinshan I and Daya Bay were given the green light, not because it would make perfect economic sense to provide nuclear energy, but rather because a combination of economic, social, technical and political considerations prompted the government to go ahead with the projects. Increasingly, development of nuclear energy is driven by market players who, at the very least, like to think their decisions are made based on the economic calculations of a project. Calculating the NPP costs becomes more important and controversial because it can indicate government policies on a wide range of issues – energy development, environmental protection and climate change, technology preferences and even labour policy.

Recent studies show that, based on the overnight cost of RMB 12000/kw (2005 prices), China would have to invest at least RMB 372 billion in initial capital investment to meet its target of 40GWe by 2020. To achieve the target of 70GWe, with 30GWe in construction by 2020, the industry would need an initial capital investment of RMB 732 billion (2005 prices). In addition, there would be another 30% of construction costs (RMB 108 billion) to be added on. By 2020, the country would need an investment of nearly one trillion yuan. This amount would double when taking into consideration inflation, interest rates, and rising labour, material and technology costs.[21]

IEA predicted that 'the cumulative investment needed to underpin the projected growth in energy supply in China is $3.7 billion (in year 2006 dollar) over the period of 2006–2030,' an average annual investment of $150 billion.[22] About 74% of this amount would go to electricity generation, transmission and distribution. To build 60–70GWe nuclear capacity by 2020, the country would have to devote more than 30% of its planned investment. The questions then are: (a) how would different energy sectors be balanced; (b) how would nuclear energy be financed at the expense of other renewable energy or electricity transmission and distribution and more importantly, (c) who can and should be allowed to invest in nuclear power plants?

Financing nuclear power projects

Nuclear energy development poses difficult economic challenges for all countries, developed and developing. NPPs are only part of an electricity system that is historically considered a natural monopoly. The firms that invest in and operate parts of an electricity system are known as public utilities. Products of a natural monopoly tend: '(1) to be capital-intensive (having sufficient fixed costs or scale of economies); (2) to be viewed as necessities (or essential to the community); (3) to be non-storable (yielding rents); and (4) to involve direct connections with customers.'[23]

Because of these features, 'much of the financing for these plants was provided by governments or with government backing or government guarantees of some kind.'[24] For example, in Britain and France, nuclear power plants were invested in and built by government-owned national utility companies, some of whose shares are publicly traded. In some other countries, such as South Korea, nuclear plant financing has evolved over time from fully government financed to fully commercial private sector financing. In Germany and the US, private sectors have often arranged commercial financing with some government credits or/and guarantees.

In China, investment in electricity has gone through fundamental changes along with its general economic reform and the global push for electricity privatisation and deregulation.[25] Before 1979, almost all revenue collected by the localities and all profits from state-owned enterprises (SOEs) were submitted to the central government while all expenditures were financed by central government transfer. Virtually, all capital investment was funnelled through downward investment planning through its five-year plans. The SPC determined the kind of resources that sectors, areas and projects would receive and 'issued annual investment control targets to government departments, which then formulated investment project plans on the basis of these targets and submitted them for approval.'[26] Funds were to be allocated according to the plan and disbursed by the Ministry of Finance through the Construction Bank, which acted as a checking mechanism to make sure the plan was followed. The SEC later gained some power over capital investment through its control of the budget for technical renovation of existing enterprises. The State Bureau of Supplies allocated materials according to the five-year and annual plan issued by the SPC.

For power projects, the central government invested in all coal-fired generation plants, large- and medium-sized hydro stations and all transmission and distribution networks, with the exception of those generating electricity exclusively for the use of enterprises, and small hydro stations, with capacity of 50MW in 1980.[27] With the reform, the central government increased its investment in the power industry, which was seen as the vanguard of general economic development. In 1981, for example, while budgets were slashed for other industries, the share of the national budget on capital construction for the electricity industry rose to 9.1%, from 6.9% in 1980.

Qinshan I was financed in this context. It had been included in the FYP of the SPC and the annual budget planning of the Ministry of Finance. There was little 'discussion and demonstration on the economic value of constructing 100 MW and 300 MW nuclear power plants in our country,'[28] or who could finance it or how it could be financed.

Policy makers in Beijing raised two questions: (a) how could the country spare the large of amount of investment, estimated to be $290 million, for a project that would not see immediate results, while it was struggling to keep the whole economy from collapsing? (b) How could the central government come up with the foreign exchanges for the necessary imports of components, such as cooling systems? The answers to these questions depended on how different ministries and bureaucracies could balance their interests, whose priorities would prevail and who could come out as a winner or loser. These had been the perennial debates on budget allocation. There was no fundamental shift in policy – the central government would allocate financial resources for the project – but the debates were intense among the leading policy makers.

On 21 June 1979, at the Plenary Session of the 2nd Session of the 5th NPC, Yu Qiuli, vice-premier and minister in charge of the SPC, delivered a report on the national economic plan. He emphasised that the investment would have to focus on agriculture, light industry, fuel and power, building materials and transport services. In particular, investment would go to those enterprises that would have quick returns with high profits and possibly with foreign exchange earnings.[29] Deng Xiaoping, although known for his support for the nuclear energy project, raised a similar concern: could the country afford it at the time.

Eventually, coal, power and transport took priority over nuclear energy. Despite these real challenges and concerns, the first NPP project went ahead. For China, the project indicated a move towards modernisation, a boost to its technical prowess, and the beginning of a major reorientation of its military-industrial complex. It was never meant to be a commercial project. Nonetheless, it turned out to be a low-budget project because its main components were from domestic sources, and its labour, land and materials were allocated through central planning.

Financing the Daya Bay project was a different matter. It pioneered utilisation of foreign investment, a combination of financing from government-budget, policy-lending, export-credits and commercial borrowing, and represented a major shift from the policies of the day. It was an experiment in seeking foreign assistance. In 1978, three delegations were sent out – one to Hong Kong and Marco, one to Yugoslavia and Romania and the third one to five countries in Western Europe (Belgium, France, Switzerland, West Germany and the UK). The lessons they brought back were that foreign capital should and could be used for development, and trade could be used for national development. Indeed, trade had already been encouraged since the mid-1970s and in the second half of 1977, 'China's imports were almost double their level in the last half of 1976.'[30] The decision on sovereign borrowing and foreign investment inflows of the central government in 1978 made it possible for the Guangdong government to propose a joint venture with Hong Kong's China Light as a co-financer and buyer of electricity.

Because of the oil crises, at the end of the 1970s, China Light was seeking to diversify its fuel mix to mitigate the impact of future oil shocks on its economy. Across the border, the Guangdong provincial government was desperately looking for investments and technologies to expand its power supply for its growing economy.

> Driven by complementary interests, China Light began exploring the joint development of a nuclear power plant in Guangdong Province. The power company and the provincial government prepared a prefeasibility study and a project proposal that they submitted to the central government for approval. ... Guangdong's proposal surprised the central government. Numerous lengthy discussions among government agencies ensued. It took two years before the State Council approved the project proposal (1982). Another two years passed before the completion of the negotiations for the project agreements ... It took an additional two years to mobilise the capital and reach financial closure.[31]

From the beginning, those in Guangdong decided that they would allow the project to pay for itself over time. The Chinese government created a corporation, the Guangdong Power Company, so that a joint venture could be formed with China Light. Guangdong Nuclear Power Joint Venture Co. (GNPJV) had its initial investment of US$400 million – $300 million was borrowed from the Bank of China and $100 million was raised by the Hong Kong Nuclear Investment Corporation (HKNIC), the vehicle for Hong Kong investors, with China Light as the major shareholder. In line with the new Chinese joint-venture practice, the joint venture would revert to wholly Chinese ownership after its expected operational life of 20 years.

The joint venture then had to raise 90% of the station's cost ($3.6 billion) to finance the project. A difficult and prolonged negotiation started with three teams of industrialists, bankers and government officials on: (a) purchasing nuclear reactors from Framatome; (b) purchasing turbines and generators from GEC and (c) designs, service and technology transfer from EDF. The estimated cost for reactors was more than US$1.37 billion, while the deal with the British was estimated at £500 million.

The Chinese team was led jointly by the party secretary of Guangdong province, Ye Xuanping (叶选平) and the vice- minister of MWREP, who was also a nuclear expert, Peng Shilu (彭士禄). The Chinese held several advantages in the negotiation: (a) there was an economic slowdown in all OECD countries and securing a deal with China would mean the creation of thousands of jobs for their countries; (b) since 1979 nuclear expansion had been severely hit by anti-nuclear movements, safety concerns, construction delays and soaring costs in developed countries; (c) western countries were competing to sell their products to China because of its future potential and (d) it would be much faster and cheaper for China to build coal-fired thermal generation plants anyway. In France, for example, some argued that by the end of the 1980s it might have as many as 15 full-scale nuclear plants in excess of requirements. It became a challenge to turn this surplus to France's advantage by increasing exports. For Britain, its turbine generators were competing fiercely with those made by the Germans and the Japanese.

The Chinese had several disadvantages: (a) China had little hard cash with which it could purchase a plant; (b) most factories in Guangdong, the fastest-growing province in China at the time, could only open three or four days a week because of the power shortage and it needed electricity and needed it immediately; and (c) the project was politically and economically important for Guangdong to build a closer relationship with Hong Kong.

It took more than two years to complete the negotiations on the three agreements. It was a very difficult negotiation partly because the Chinese demanded a 20% discount for the French reactors and a 25% discount for the British turbine generators. They also demanded the French offer export-credit terms comparable to the Japanese Exim-bank. France was bound by the OECD's 'gentleman's agreement' that set the interest rates for export credits in the range of 10–12.4%, while Japan was allowed to set its rates at 0.3% over its long-term prime rate, which was about 5.5%.[32] Meanwhile, the ultimate objective for many in China was to build their own NPPs and consequently link the deal to the technology transfer.

When the contracts were signed, it was a great relief for all sides. The total value of the plant was about US$4.1 billion (£2.8 billion). The French share, including supervision of design and construction by EDF, was about 10 billion French francs (£920 million), with 60% going to Framatome. Even at 20% less than the asked-for price, GEC secured an initial order of £250 million to supply equipment for the Day Bay project. The order was significant for GEC because it would help secure about 7000 jobs for about four years from 1987, and would allow GEC to compete with Mitsubishi of Japan for making large turbine generators. 'Obviously, this first order from China is a very important milestone in our efforts to enter the Chinese market,' explained the head of GEC Turbine Generators.[33] This also represented the mentality of the French. The Chinese were apparently pleased with their bargaining: 'We've been able to reduce the equipment costs from US$1.5 billion to US$1.1 billion,' explained Peng Shilu. 'We allow foreign companies to make some money, but not too much.'[34]

The deal with the British did not involve aid or soft credit terms. Rather, the Bank of China borrowed money on behalf of the Guangdong Nuclear Power Joint Venture Company. This loan was put together with the 'extremely helpful and cooperative' British government, by a consortium of 10 British banks, led by the Midland Bank, and the loan was underwritten by the UK's Export Credit Guarantee Department (ECGD). The loan, covering 85%

of the contract values, was set for payment in 15 years and at an interest rate of 9.85%, lower than the rates in the market, especially lower than the US rate. The remaining 15% of the GEC contract value would be borrowed by the Chinese at competitive commercial rates. Banque Nationale de Paris was doing the same with French banks for the Framatome contract and the loans were guaranteed by Coface, the ECGD's French counterpart.

Diversified investment in power projects

Utilisation of export credits became a trait of NPP construction in China and financing Daya Bay pioneered the early diversification of investment. It mirrored new thinking and new ideas – lowering the entry levels would help attract investment in a sector where investment was intensive and at a time when the central government was no longer in a position to finance. This change was the major contributor to a rapid expansion of power generation capacity in China and had long-term impacts on future investment. In 1985, the central government 'introduced the Provisional Regulations on Encouraging Fundraising for Power Construction and Introducing a Multi-Rate Power Tariff,'[35] which removed the government monopoly as the sole investor in electricity (see Table 5.8). The contribution from the central government budget to the power sector financing dropped from 60% in 1980 to 0.2% in 1996 and foreign funds jumped from nothing to nearly 12%. Provincial and local governments also mobilized their own resources by forming partnerships with the central government, SOEs or foreign companies to acquire necessary capital for power generation facility expansion.[36]

Table 5.8 Power sector investment sources for capital construction in % (1980–96)

	1980	1985	1991	1996
Government appropriation	66.4	41.2	0.5	0.2
Operational funds			7.0	1.6
Domestic loans	25.0	24.0	23.6	38.2
Foreign funds		5.0	10.9	11.7
Transferred to local authority			3.0	1.5
Fund raising		9.0	18.5	16.9
Bonds			7.4	1.0
Self-financing	9.6	6.4	15.7	24.8
Oil to coal fund			4.6	1.17
Other		14.4	7.7	2.8
All financing (billion yuan)	4.124	9.669	31.601	97.419

Source: World Bank, 'The Private Sector and Power Generation in China', World Bank Discussion Paper No.406, February 2000, p.36.

The policy of diversified investment sources led to an expansion of investment in the power sector and the investment in electricity generation as a share of the total investment in energy industry increased steadily in the 1980s, especially the second half of the decade, from 4.01% in 1980 to 7.70% in 1984 to 33.01% in 1990.[37]

The policy of diversifying investment resources also created an opportunity for Guangdong to build its second NPP after the Daya Bay project. In the early 1990s, the Guangdong government proposed another NPP next to Daya Bay. The central government could not and did not want to spare the resources for a single project that would only be beneficial to a single province. China Light that had financed 25% of the initial investment in Daya Bay declined to participate in the new project. When the feasibility study was under way, questions were raised on how to finance it. Li Peng, the Premier of the State Council, who had supported the Daya Bay project, acknowledged the benefits of the project in public, but refused to endorse the new project because of the macroeconomic situation in the country.

As soon as the State Council approved the establishment of the China Guangdong Nuclear Power Corporation (CGNPC) in 1994 based on the structure of the GNPJV, CGNPC announced its strategy of 'rolling development' – using the profits from Daya Bay to finance another NPP at Lingao. It would do so with or without the central government's support. CNNC lent its support, as its chairman, Jiang Xinxiong (蔣心雄), stated that the income from Daya Bay would 'provide the necessary capital for the construction of a second nuclear project in the province.'[38] There was no mention of the initial guaranteed preferential loans from the central government. Many questioned whether the country could afford another major turnkey nuclear project or if it was a good time to invest in such an expensive project when the economy had gone through regular inflation cycles. All these concerns were brushed off by the Guangdong government because 'the emperor is far away' and by CGNPC because it came up with the finance. Money talked.

Rules are developed to achieve certain defined objectives and they shape the opportunities in a society where each individual responds differently. New rules and their enforcement often produce positive and negative, expected and unanticipated consequences. Devolving investment authority brought about a rapid expansion of generation capacity – total installed generation capacity in the country almost tripled between 1987 and 1995 – and an increase in power production. For example, 'Guangdong province spent 8.28 billion yuan ($1.36 billion) from 1986–1990 to add 44 000 megawatts of capacity to its grid, more than the total installed capacity achieved in the 36 years before 1986.'[39] The installed capacity quadrupled from 1987 to 1993 and then doubled again from 1993 to 1998.

As 'much of the financing burden shifted from the central government to the provinces and cities,'[40] new problems emerged as well, with the implementation of new rules – inadequate investment in transmission and distribution networks, an increasing number of small-sized inefficient power generation units, competition for ownership of new power plants, and their off-shoot problems, such as unreliable and unstable power supply, high line loss, and impediment to competition. Changed rules also fostered large enterprises whose behaviour became increasingly difficult for the central government to regulate. Even though nuclear power projects must be approved by the State Council, having one of the two nuclear power companies in the province gives Guangdong substantial power and economic influence in the industry.

Who invests in nuclear power plants?

The nuclear industry was dominated by two investors – CNNC and CGNPC – and these were the only entities that were allowed to be controlling shareholders even after the country conducted serious commercialisation and corporatisation in energy industries in the mid-1990s. This monopolistic structure follows the normal international practice – one large state-owned corporation owns and operates all NPPs in most countries. China had two major players because of the historical development of its civilian nuclear programme. CNNC was the owner and operator of Qinshan I. GNPJVC was the owner and operator of the Daya Bay project. Its capital came from China Light (25%), the MNI and then the CNNC (33.75%), Guangdong province (33.75%) and the MEP and then SPCC (7.5%). While CNNC controlled nuclear fuel production and services and basic research in the nuclear industry, CGNPC was the one that grew very quickly.

This structure was challenged in the 2000s, partly because the expansion of a nuclear energy programme requires large amounts of financial resources. Investment in NPPs before the mid-1990s depended on government-guaranteed borrowing from domestic and international sources when China did not have financial markets that would allow investors, public or private, to raise the capital. Even after the two stock markets in Shanghai and Shenzhen went into operation in the mid-1990s, nuclear companies were not allowed to raise capital there. CNNC was loss-making until 2003 when it finally made a profit of about RMB 240 million. In the following five years, it significantly strengthened its financial position. In 2008, its estimated profit was RMB 4.8 to 7 billion, while that of CGNPC was about RMB 4 billion.[41] Alone they were clearly unable to meet future demand. In November 2005, the National Energy Leading Group decided to grant the China Power Investment Corporation (CPI) permission to invest in a nuclear power project as a controlling shareholder, partly because an expansion of a nuclear energy programme demanded more financial resources and partly because CPI had already controlled some portions of NPPs in operation.

CPI was created in 1995 by the MEP when the State Council decided to corporatise the business segments of the Ministry into the SPCC. CPI, whose assets were spin-offs from the Ministry, was to carry out several responsibilities, described by its first president as 'floating public power plants assets, issuing corporate bonds, establishing power development funds and channelling foreign investment for build-operate-transfer (BOT) power projects.'[42]

CPI's international financing arm, China Power International, was designed to raise funds on the overseas market. The SPCC hoped that through CPI it would eventually be able to list in Hong Kong, New York and London – following the path trodden by China Telecom in raising funds for the country's telecommunication requirements. Quickly, CPI gained international recognition in raising not only domestic but also international funds with its active involvement in the power industry's development. It also obtained 'stakes in nine plants, though not all the assets of it [held] in these plants have been transferred to its books, pending financial agreement by local power officials and other parties on asset valuation.'[43]

In 2002, in another major round of reform, China moved from a single, vertically integrated utility to two grid companies (a large one covering most of the country and a small one in the south) and a diverse set of generation companies (five large companies that were spin-offs from the original incumbent and a large number of other companies). The five companies are China Huaneng Power Group Corp, China Datang Corp (大唐), China Huadian Corp (华电), China Guodian Corp (国电) and China Power Investment Corp (CPI, 中电投). They joined the elite team of large state-owned corporations under the supervision of the State-owned Asset Supervision and Administration Commission (SASAC). This unbundling was made on the principle that none of the five new generating companies were allocated more than 20% of the shares of any one of the regional power markets. Each was allotted about 20GW generation capacity. By the end of 2008, they controlled 44.9% of the country's total generation capacity (353GW in total).[44] These changes took place in the context of wider economic reforms to promote growth and economic development in China and also in the context of a global shift in organising the power industry.[45]

When the SPCC was unbundled in 2002, CPI, the smallest of the five companies, absorbed some non-productive segments of the SPCC. It also inherited the stakes in all NPPs in China, originally held by the MEP/ SPCC (6% of Qinshan II, 20% of Qinshan III, 7.5% of Daya Bay and 10% of Lingao). The current registered capital of CPI amounts to RMB12 billion and it owns a little over 8% of the country's total installed generation capacity, far behind the other four majors, but it does own the nuclear generation capacity that the five major generation companies would like to have. It owns a total generation capacity of 51GW, among which 15.54GW is hydro (20.3%), 41.12GW is thermal (79.1%), 0.33GW is wind (0.6%) and 1.35GWe is nuclear (3.37%).

With the changed government policy, CPI invested in the Haiyang NNP in Shandong (6x1000MW), as the controlling shareholder, with 40% of the ownership and the rest is split between CNNC (20%), Guodian (20%), Shandong Luneng (10%), Huaneng (5%) and Yantai Power Development Corporation (5%). The first stage of the development has two units of AP1000 with a total investment of RMB26 billion. The construction, according to IAEA, officially started on 24 September 2009, and CPI hopes that it would be commissioned in operation by 2014. Only time will tell, because in 2007 some sections of the media reported that the Haiyang project would go into commercial operation in 2010. The public debate on whether there should

Table 5.9   Financial structure of nuclear power projects in China

	Capacity	Equity/total investment	Ownership structure	Electricity market
Daya Bay 大亚湾	2x1000MW	US$400 million/ US$4 billion	China Light (25%) GNPJV (75%): Guangdong (45%), CNNC (45%), SPCC (10%)	70% Hong Kong 30% Guangdong
Lingao 岭澳	2x1000MW	US$400 million/ US$4.2 billion	CGNPC (100%): Guangdong (45%), CNNC (45%), SPCC (10%)	Guangdong
Qinshan I 秦山 I	1x300MW	RMB2.4 billion	CNNC (100%): CNNC (50%), Zhejiang (20%), Shanghai (12%), Jiangsu (10%), Huadong (6%), Anhuai (2%)	Zhejiang Huadong
Qinshan II 秦山 II	2x600MW	US$1.968 billion	CNNC (50%), Local (50%)	Zhejiang Huadong
Qinshan III 秦山 III	2x728MW	RMB1 billion/ US$2.88 billion	CNNC (51%), CPI (20%) Local (29%)	Zhejiang Huadong
Tianwan 田湾	2x1000MW	$3.2 billion	CNNC (50%), CPI (30%) Jiangsu Power (20%)	Jiangsu Huadong
Hongyanhe 红沿河	4x1000MW		CGNPC (45%), CPI (45%) Dalian (10%)	Liaoning
Ningde 宁德	4x1000MW	RMB 49 billion	CGNPC (51%), Datang (49%)	Fujian

Source: 高阳，邹树梁，'我国核电产业的垄断性分析及规划改革模式邹议' 南华大学学报，(Gao Yang and Zou Shuliang, 'The Monopolisation of China's Nuclear Energy Industry and the Plans for its Reform', *Nanhua University Press*, 6:6, December 2005, pp.27–28.

be a nuclear power plant built in the region along the coast contributed greatly to the delay in the project.

Since 2005, CNNC, CGNPC and CPI consequently are the only controlling shareholders of nuclear power plants in China (see Table 5.9).

After the unbundling of SPCC in 2002, the generating companies were set free and 'quickly learnt how to benefit from a situation in which they are no longer part of "a plan" but are not yet under the effective supervision of a regulator ... they find it more lucrative to grow market share and increase profits, and far less interesting to cut costs and become more efficient.'[46] In less than four years, they significantly expanded an amount of capacity ranging between 30 and 38GW, 150–200% of their initial allocation. By the end of 2008, together they controlled more than half of the country's generation capacity (see Table 5.10).

Since 2003, their capital investment raised constant concerns because of the overheated economy and potential oversupplies of generation capacity.

Table 5.10 Generation capacity and total assets of the five generating companies in China, 2002–08

		2002	2003	2004	2005	2006	2007	2008
Huaneng	Generation capacity (GW)	26.8	31.7	33.6	43.2	57.2	71.6	85.9
	Total assets (billion yuan)	125	146	156	227	286	376	464
Guodian	Generation capacity (GW)	22.1	25.3	29.3	35.1	44.5	60.1	70
	Total assets (billion yuan)	67	75	101	132	193	246	304
Huadian	Generation capacity (GW)	25.5	28.6	30.8	38.8	50	63	69
	Total assets (billion yuan)	84	96	118	147	198	243	293
Datang	Generation capacity (GW)	23.9	27.5	33.5	41.7	54.1	64.8	82.4
	Total assets (billion yuan)	94	112	140	183	226	295	408
CPI	Generation capacity (GW)	–	23	24.4	29.5	35.5	43	52
	Total assets (billion yuan)	–	88	103	138	178	218	275

Sources: Data is from the website of each of the five generating companies: http://www.chng.com.cn; http://www.chd.com.cn; http://www.cgdc.com.cn; http://www.china-cdt.com; http://www.cpicorp.com.cn, accessed on 20 December 2009.

Table 5.11 Composition of generation capacity of five generating companies, 2008

	Huaneng	Huadian	Guodian	Datang	CPI
Total (GW)	86	69	78	90	52
Thermal (%)	92.5	86.8	88.9	80.8	79
Hydro (%)	6	12.7	6.5	16.6	20.3
Wind (%)	1.2	0.5	4.54	2.61	0.6

Source: Data is from the website of each of the five generating companies: http://www.chng.com.cn; http://www.chd.com.cn; http://www.cgdc.com.cn; http://www.china-cdt.com; http://www.cpicorp.com.cn, accessed on 20 December 2009.

'The central government has been trying to restrict new investment and consolidate existing capacity...with very few visible signs of success.'[47] In addition to their 'empire-building' mentality, and despite its high risks and uncertainties, these large State-owned generating companies were motivated to stake a claim in the nuclear industry by utilising the changes in government policies. The adoption of the Medium- to Long-Term Nuclear Energy Development Plan (2005–20) presented new opportunities for their expansion. Another motivation behind their push to become involved in the nuclear sector was a heavy reliance on thermal capacity.

Their heavy dependence on thermal generation capacity became more problematic after 2002 (see Table 5.11). With rising electricity demands, power companies had difficulties in getting sufficient coal, partly because of tight coal supplies and partly because of rail transportation congestion. Coal prices had long been deregulated and were, by and large, set by markets. The coal prices rose 40% in 2004 and the trend continued, and power companies could not pass on the rising costs to end-users as power tariffs are set and regulated by the central government.

As large state-owned corporations, the five power-generating companies had obligations to supply electricity even as a loss-making exercise. In 2008, the five together incurred a total loss of RMB 32.5 billion. Their losses from thermal generation were more than RMB 40 billion. The debt–equity ratio for all five power companies was more than 80% and Huadian had the highest debt–equity ratio of 87.6%. The profit margin of Huaneng dropped from a positive RMB 10 billion in 2007 to a negative loss of RMB 5.8 billion in 2008. High dependence always means vulnerability and these power companies wanted to change this. Finally, their large thermal generation capacity brought increasing pressure to reduce GHG emissions as the central government issued top-down targets for energy efficiency.[48] Nuclear was now seen an alternative for them in dealing with greenhouse gas emissions.

The five power-generating companies lobbied the central government to lower entry barriers to the nuclear energy sector and they certainly made their voice heard. The media were used to spread the message that the government should break the nuclear monopoly and allow them to 'compete' in this sector. Facing pressure to give up their monopoly, CNNC and CGNPC behaved in the same way as other monopolies or oligopolies in the capitalist system did, trying to hold on their position by lobbying the government. The difference between these Chinese monopolies and their international counterparts was that they had closer ties and easier access to government agencies in support of their position.

In 2005, the National Nuclear Safety Administration (NNSA) organised CNNC, CGNPC and CPI to draft a set of requirements for entering the nuclear power industry. The draft of the regulation, though not at that time approved by the State Council or the NPC, was already considered binding by all power companies. To no one's surprise the requirements for entering the nuclear energy sector were written in such a way to protect these very oligopolies that others tried to challenge. The document contained three major requirements for any company to become a controlling shareholder of a new nuclear power project:

- 20% of shares in two separate nuclear power projects, one of which had to be the second-largest shareholder.
- At least six years experience in managing nuclear power projects.
- At least six years experience of operating nuclear reactors.

These requirements meant exclusion, rather than inclusion, for other investors, except the three that participated in drafting the requirements. Indeed, currently, only CNNC and CGNPC meet these requirements. CPI is expanding its ownership of nuclear power plants but has not demonstrated any desire to become an operator. The other four major power-generating companies are struggling to get their foot in the door and are part of the recent drive for nuclear expansion into interior provinces.

In 2006, Datang joined forces with CGNPC and formed the Fujian Ningde Nuclear Power Company Limited, with CGNPC controlling 51% of the shares and Datang 49%. This is a 4x1000MW project and its initial investment was 49.4 billion yuan. In 2008, with the support of Datang and CGNPC, the Ningde project secured 40 billion yuan from the China Industrial and Commercial Bank and another three policy banks – the Agriculture Bank, Development Bank and Bank of China. Its main finance was from domestic sources because the State Council made it clear that foreign borrowing for nuclear power projects should be restricted. Huadian joined forces with CNNC and the Fujian Provincial Government in the Fujian Fuqing nuclear project (6x1000MW) and work for the first stage (2x1000MW) began in November 2008 and June 2009. CNNC is the controlling shareholder (51%), Huadian owns 39% of the shares and the rest was contributed by the Fujian Development Corporation, a provincial government investment instrument.

Huaneng is the largest and oldest among the five generating companies, and the only generating company listed as 'China's Backbone Corporation'. About 95% of its total generation capacity of 89GW is thermal. For some time, Huaneng had harboured a strong desire to move into the nuclear arena. In March 2004, Huaneng signed an agreement with Tsinghua University and China Nuclear Engineering and Construction Corp (CNEC) to build an experimental module of HTR-PM in Rongcheng, Shandong. In November 2005, Huaneng created a subsidiary, Huaneng Nuclear Development Company Ltd, to invest in the nuclear sector. CNEC is the only company able to install the components and systems of the nuclear island in China. INET of Tsinghua University would be responsible for the engineering design work for the HTR-PM. The site for this demonstration HTR-PM is in Shandong.

In December 2006, the three parties formed a joint venture with Huaneng as the controlling shareholder (47.5%), CNEC (32.5%) and Tsinghua University (20%). The initial investment was 3.2 billion yuan: 1.2 billion yuan came from the government budget allocation, 2.4 billion yuan was a guaranteed loan for technical assistance, and the rest of the 600 million yuan came from the three shareholders. Huaneng also invested 5% of its share of Haiyang in Shandong. The high risks involved in the HTR-PM project made many at Huaneng wonder whether it was a wise decision to take on the project.[49] Huaneng is known to lobby harder than other companies in staking its claim in nuclear energy development.

Finally, in the early 1990s, as an experiment to see how the core of the capitalist system would work in China, the central government decided to create two stock markets – the Shanghai Securities Exchange (December 1990) and the Shenzhen Stock Exchange (July 1991). In the mid-1990s, to facilitate the reform of SOEs, the government expanded the two stock exchanges so that enterprises, state-owned or otherwise, could issue corporate bonds to invest in their expansion. These changes paved the way for nuclear energy development in the 2000s.

After years of relying on government budget allocation and/or bank loans, in August 2007, China launched a key reform to allow the China Securities Regulatory Commission (CSRC) to take over authority from the NDRC to approve listed companies' issuance of corporate bonds of one year or more. CNNC and CGNPC responded quickly. After two successful issuing of bonds, on 11 November 2008, CNNC raised another 1.8 billion yuan on a five-year bond and on 15 July 2009, it raised 4 billion yuan on a ten-year bond to finance the Fuqing project exclusively. In early 2009, CGNPC issued bonds amounting to 7 billion yuan to finance Yangjiang and Taishan projects in the province. The bond market allowed the companies to raise huge capital to finance nuclear energy expansion. It also created opportunities for corruption because of the lack of regulation. This was the triggering point for the downfall of the general manager of CNNC (康日新) in 2009.[50]

One problem with the current investment structure in NPP projects in China is that all the players are state-owned under the direct control of the SASAC. This means that 'many of the risks associated with construction costs, operating performance, fuel price changes, and other factors were borne by consumers rather than the supplier'[51] through lending by policy banks. State-ownership or government-guaranteed loans for nuclear development are often needed as the American government has recently 'enacted measures to provide loan guarantees and tax incentives for...new NPPs, intended to overcome investor reluctance to take on first-of-a-kind risks.'[52]

Governments everywhere have adopted similar policies. They have a series of instruments at their disposal to 'assist' nuclear power project financing. Such guaranteed assistance can also create moral hazards as state-owned corporations invest recklessly for their empire-building. This now seems to be the case in China. Three major players are joining forces with provinces, planning as many as 70 sites for potential nuclear power plants. Whether the projects go ahead or not, as soon as a site is identified and a company is formed with one of the three nuclear companies, the province could start putting money into infrastructure: roads, water and electricity, all of which create jobs for the local economy and which also became the major contributors to the heated economy in 2009–10.

How should electricity be priced?

How to price electricity can either facilitate or obstruct NPP development. 'The price of a unit depends not only on the price of any fuel involved, but on asset accounting, taxation, regulation, risk, subsidies, network and system effects, and other factors usually unmentioned.'[53] Any price given to electricity is arbitrary, and is decided by contracts for services on who owns and operates the infrastructure, who has access to it, and who uses it and for what purpose. Therefore, power tariff setting is often as much a political matter as it is an economic matter. It is frequently regulated, but not set, by the government regulatory agency because of the natural monopoly nature of electricity, especially its transmission and distribution.

In China, power tariff setting has gone through several rounds of changes in the past 30 years. In 1977–84, the government raised prices for all essential commodities to encourage production. 'Negotiated' prices started to appear in 'free markets' where farmers were allowed to sell some of their produce. In the second period (1984–88), the central government maintained price setting for many commodities, while also adopting 'guided-pricing' for many more. A dual-pricing system was also adopted – those commodities under the central plan could be sold only at a government-set price, while any extra (out-of-plan) portions could be sold at a negotiated price between buyers and sellers.

Major reform measures were adopted in the mid-1980s. 'To alleviate the extreme shortage of electricity and encourage more generation from different sources, the state in 1985 implemented a "diversified tariff system for electricity produced outside the plan".'[54] The dual-pricing system had two components: planned output from the existing power generation plants continued to be distributed at prices fixed by the government and to the end-users as designated, whereas the extra supplies could be charged at a price 20% higher than the planned one. In the same year, the central government allowed the power industry to pass on some of the increasing costs of fuels and transportation to end-users. This was adopted to help the power sector deal with the squeeze between ever-rising costs for fuels and transportation and the government-fixed power tariffs. When 'fuel costs accounted for an average of 50–55% of power industry operating costs,' the policy alleviated some pressures from the power sector.[55] In 1987, the State Council issued the Price Administration Regulations, which prescribed the three categories of pricing: *market-determined, government-guided* and *government-set* prices. This freed some of the prices but not that of electricity, which remained under the category of government-set prices. Reforms, nonetheless, moved on. In 1988, the central government adopted three policies that changed the power tariff setting significantly:

1. It changed the policy from SOEs submitting profits to paying taxes and it 'raised tax on production and sale of electric power from 15% to 25%, including a 10 yuan/MWh charge on producers and a 10% revenue tax on distributors.'[56]
2. It began levying fees for developing selected generation and grid projects. These included: (a) a surcharge of 2 cents/kwh to be used for substituting coal for oil in the early 1980s, which was soon used to finance the formation of Huaneng Power Corporation; (b) a surcharge of 2 cents/kwh to finance local power development (in 1996, the central government divided this surcharge into two parts: 1 cent/kwh for local power development and 1 cent/kwh for transmission construction) and (c) a fee of 4 cents/kwh was levied to help finance the construction of the Three Gorges Dam, which was later increased to 0.007 yuan/kwh.[57]
3. It started to allow investment from other sources and new plants were allowed to charge higher prices to recover costs and to provide a fixed return on profit. This created a two-tiered pricing system – 'new price for new plant'. The aim was to encourage investment from non-government sources.[58]

Initially, the 'new price for new plant' policy was only applied to plants constructed between 1986 and 1992 that did not use central government funds. The principle was incorporated into the Electric Power Law of 1995, which stipulated that tariffs should permit cost recovery with allowance for taxes and reasonable profits. The policy of 'cost-profit-tax' was designed to encourage domestic and international investors other than the central government to invest in power generation. The formal power tariff approval consisted of a two-tier system: the establishment of initial tariff levels and annual tariff adjustments. The annual tariff review process was particularly adopted to ensure that increases were 'socially' acceptable. In the case of high tariffs, companies were offered financial incentives to keep the tariff increases at acceptable levels.

The 'new plant, new price' regulations played a major role in attracting domestic and foreign investors to China's power sector and in alleviating power shortages. Meanwhile, the policy of 'cost-plus pricing' created a situation of one price for one plant, one price for one region and one price for one category of end-users. It created opportunities for various levels of government to add on a wide range of legal and illegal surcharges and fees. Throughout the 1990s, the government tried to unify the power prices and clean up the illegal surcharges and fees, which had run up to 21.7 billion yuan by 1998.[59] It then introduced a policy of 'operating period tariff', which set the tariff on the expected lifetime of the plant, rather than on the debt payment period. The objective was to deal with the opaque power tariff setting and to control and lower the capital cost of new plants and to place the responsibility for negotiating suitable financing terms on the project sponsors. In 2004, another new pricing policy was adopted: 'each price is based on current estimated construction and operating costs of the various technologies, specific to the provinces in which they are located.'[60] This policy was to encourage the construction of coal-fired generation with flue gas desulphurisation. The price set based on desulphurisation has been used as a base for comparison of thermal and nuclear power.

These measures fundamentally altered the previously planned system by changing the incentive structures for producers. Unfortunately, these reform measures carried with them some serious problems – inflation became a recurrent trend and corruption spread quickly. The combination of these two problems triggered student protests in 1989, and resulted in a tightening up of investment policies, which prevailed during the third period (1989–92). From 1993, price reform policies were speeded up. The price for most commodities, including coal, was open to the market operation. One exception is electricity.

Nuclear power is often the price-taker, especially when it contributes only a small proportion of electricity to the grid. For nuclear energy, power tariffs were set retroactively after the projects were completed, rather than according to the contracts for the power projects, as was the case in coal-fired thermal plants. In addition, 'one price for one plant, one price for one region' was the principle by which power tariffs were set. Initially, power tariff setting for nuclear power also followed the principle of guaranteeing the recovery of cost. This principle was abandoned partly because initial capital investment in each nuclear power project varied greatly: some was from direct government budget allocation; others were from domestic or foreign borrowing; and some were in the form of an equity-debt arrangement secured from the market. In sum, nuclear power tariff setting now remains an *ad hoc* exercise. For example, for Daya Bay, the principle was to recover the value of its investment. Only in the following years were profits adjusted into the price setting.

For the Lingao project, initially the power tariff was based on a combination of the operating period of the nuclear station, 80% of the load factor, 10% profit margin and depreciation of 25 years based on the initial capital investment (equity portion, not including debt). This arrangement was set in the initial contract between the operating company (CGNPC) and the China Southern Grid for five years. Before the five-year period had ended,

Table 5.12 Comparative power tariffs in a given grid, 2005 (yuan/kwh)

	Average price to get on grid	Benchmark price for coal-fired	Nuclear power price		Differential from average benchmark	Differential from coal-fired price
Guangdong	0.485	0.45332	Average	0.415	−0.070	−0.0382
			Daya Bay	0.414	−0.071	−0.0392
			Lingao	0.429	−0.056	−0.0242
Zhejiang	0.441	0.4195	Average	0.426	−0.015	+0.0065
			Qinshan I	0.420	−0.021	+0.0005
			Qinshan II	0.393	−0.048	−0.0265
			Qinshan III	0.464	+0.023	+0.0445

Note: Power tariffs refer to the price generating companies receive for their product on an output basis, i.e., per kilowatt hour. In a market system, they will usually be sufficiently high to recover their annual capital, fuel and operation costs.

Source: 刘树杰, 杨娟, 陈扬, 彭苏颖, '核电价格形成机制研究', 中国物价 (Liu Shujie, Yang Jun, Peng Suying, 'An Analysis of Nuclear Power Pricing', *Chian Price*), October 2006, p.19.

renegotiation on this price arrangement took place to develop some new power tariff setting for Lingao.[61]

'One price for one plant, and one price for one region' created a similarly chaotic power tariff for nuclear as it did for thermal. In 2005, for example, the nuclear power tariff was 0.414 yuan/kwh for Daya Bay, 0.429 yuan/kwh for Lingao, 0.40 yuan/kwh for Qinshan I, 0.393 yuan/kwh for Qinshan II and 0.464 yuan/kwh for Qinshan III.

In 2004, NDRC readjusted tariffs for different types of generators, the 2005 figure being:

- Nuclear operators had an average tariff of around 0.41 yuan/kwh.
- Thermal generators had an average tariff of around 0.36 yuan/kwh.
- Hydro generators had an average tariff of around 0.30 yuan/kwh.[62]

Table 5.12 shows the difference not so much because tariffs were decided with consideration of investment costs, rather it was set based on the historical cost of the region with thermal power as the benchmark. This explains why nuclear power could currently be sold at a price equal to or cheaper than electricity generated using other technologies. These arbitrary prices could hardly be seen as the corresponding reflection of the combined costs of investment, operation and management, fuel, discount rates and risks. Many have argued that because the price for thermal power in China is below the 'cost-plus marginal profit', using the power tariffs for thermal generation for nuclear power pricing is clearly a distorted price setting.

Several developments can explain the difficulties in power tariffs at a level that would be conducive to efficiency and fair to customers as well as investors.

First, in most countries, the public is confused about whether nuclear power is a cheap source of electricity because 'when most existing nuclear power plants were planned and constructed, electricity supply was the responsibility of price-regulated or state-owned, vertically integrated utilities, often with a monopoly or near-monopoly in their national or regional service areas.'[63] This means that power tariffs were related to costs on average and utilities were not under competitive pressures even when they also owned and operated nuclear power stations. This can explain the practice of setting nuclear power tariffs by using thermal as a benchmark, and also its consequent low prices.

Second, SERC acknowledges the price for generators completed before 2006 and for all renewables, including nuclear, was low.[64] It is estimated that average power prices are 10–15% below long-term marginal costs in coastal provinces and 30% and more in interior provinces.[65] Low power tariffs are often blamed for inadequate investment and low efficiency in the electricity industry.[66] Yet, power tariffs are often decided not only based on economic calculations but also for social and political reasons. Electricity is not a normal commodity and its price has far-reaching impacts on the country's general economy and standard of living.

Even though 'cost-reflective prices across the value chain would provide signals to trigger efficient investment and to curb consumption'[67], it has never been easy anywhere to set power tariffs according to this principle. Radical changes to power tariffs are unlikely anywhere in the world. Given that power tariffs started at a low rate in China, it is even more difficult to change them. In 2003, for example, a study done jointly between the China Academy of Environmental Sciences and Tsinghua University showed that the cost of SO_2 emissions would run as high as 110 billion yuan a year and this would and should be translated into 0.0676 yuan for every kilowatt hour of electricity generated.

In the same year, a new regulation on waste management was adopted jointly by the State Planning Commission, the Ministry of Finance, the State Environmental Protection Administration, and the State Economic and Trade Commission. The regulation included 0.63 yuan for every tonne of SO_2 emission, which meant that a power plant with a capacity of 1000MW would have to pay 47.25 million yuan a year. Despite this calculation, however, the regulation demanded only 0.0086 yuan/kwh be added to the base price (1.3% of that suggested). When power plants had to pay only a little over 1% of the real costs they incurred in SO_2 emission, it is not surprising to see more than one-third of the coal-fired thermal power plants have done nothing with desulphurisation.[68]

Third, nuclear energy is developed everywhere with considerable government assistance, either with guaranteed and preferential loans, or outright subsidies. This is partly because the initial capital investment costs are high. More important, it is because nuclear energy is seen as an embodiment of: (a) advanced technology development; (b) economic sophistication;

(c) national defence capability; and increasingly (d) the ability to secure energy supply and (e) the ability to tackle climate change. These are what economist like to call 'externalities' that can hardly be measured in pure dollar terms.

In China, preferential financial assistance is provided to nuclear energy development by the central as well as provincial governments. The assistance comes in all forms – direct subsidies, guaranteed loans and direct investment. For example, CNNC and CGNPC are state-owned. As discussed in the chapter on politics (Chapter 4), they are not required to pay dividends to the ultimate shareholder through either SASAC or the Ministry of Finance. And neither are they fully expected to recover the costs of their initial investment in a short period. When a substantial share of initial investment comes from loans made by policy banks, the interest rates and terms of these loans can be quite generous. As many economists have discussed, income tax can be a major factor affecting the nuclear power tariffs and financial performance of nuclear companies. It is not included in the calculation of power tariff setting in China.

Fourth, in most countries, research and development on nuclear technology is financed by government and companies pay for their application. Transfer of technology is often free. Since capital investment costs constitute the largest share of generation costs (60%) and 'capital cost savings obtained by standardisation and series construction are reported to range from 15% to 40%,'[69] a large share of the capital investment at the moment is to 'purchase' new technology for future standardisation. Instead of allocating a large amount of the budget to basic research and development, financial resources are used to purchase the most advanced technology, which is used as a reference to develop domestic technology and industry. It is difficult for the government to demand this cost be recovered through electricity pricing. This issue is discussed in the chapter on technology transfer (Chapter 6). If China expands its nuclear generation capacity to 40–60GWe, it will create 500–700 billion yuan in business by 2020.

Fifth, the electricity industry retains all the features of natural monopoly. Therefore, even after market competition is introduced into generation, regulations should be in place – regulation of prices, investment and service qualities. In 2002, when the central government unbundled the SPCC, it created SERC. The regulatory responsibilities, however, are split among a number of different organisations and the regulatory powers of each of them is lacking in clear definition. NDRC is responsible for planning and price regulation. The Ministry of Finance has some decision-making powers relating to certain financial rules and cost standards; and the SASAC exercises a supervisory role over state-owned enterprises, in particularly in appointing and supervising senior executives.

As NDRC insists that power tariffs are one of a few instruments it still has to manage macroeconomic stability, SERC was not given sufficient authority in regulating pricing. Many have seen this as an exercise of NDRC to hold on its control over an industry where market forces are playing an increasingly important role. The debate continues.

Conclusion

High initial investment in nuclear power plants has always been a challenge for all countries wishing to expand nuclear energy programmes. The economics of nuclear power relative to fossil-fuelled generation, particularly coal, has improved with carbon pricing and rising coal prices in China. The following questions remain: Who should finance nuclear expansion in China? How can the government balance the allocation of scarce resources when other social and economic demands remain high? What would be the incentive structures to attract investors in a sector involving many political, economic, technical or security risks?

China was one of the first among developing countries to use foreign capital for nuclear energy development in its early stages. This was made possible when the central government abandoned the old isolationist economic policies to allow sovereign borrowing and foreign investment at the end of the 1970s. The second major reform was the diversification in electricity investment that ensured a rapid expansion of power generation. It also shaped the incentive structure within which provinces, large generation companies and the traditional nuclear power companies all now try to invest in nuclear energy expansion. Financing new nuclear power projects remains a challenge because of its huge up-front investment. The government in one form or another has been either a direct investor or facilitated other sources of capital for all existing power projects. This practice is increasingly at odds with present strong market liberalisation policies.

Moreover, China remains a developing country. Demands for governmental resources are multiple and high. It is contentious to 'subsidise' nuclear power plants while other social and economic demands are not met. The huge capital requirements, combined with risks of cost overruns and regulatory uncertainties, make other investors and lenders very cautious even when demand growth is robust. Currently, in China, the government remains the last resort of the risk taker. And this distorted investment environment has created a situation where utilities want to get into the nuclear sector regardless of risks.

Cost-reflective pricing is the key to ensuring adequate investment. Indeed, getting pricing right can go a long way towards both the economic and energy efficiency. In China, the price of nuclear power does not yet have a life of its own. It is made with coal-fired thermal as a reference point. It is still difficult to calculate the real 'costs-plus profit' for nuclear power because a large amount of capital investment comes from the government sources in one form or another. Many hidden subsidies remain. The overlap in regulatory functions between the NDRC and SERC further undermines the government capacity in policy making, investment approval and in pricing regulation.

The IEA has suggested the Chinese government adopt 'a more transparent approach to pricing and the application of cost – reflective methodologies are needed to identify the extent of the use of public funds in the power sector – and to wean the power sector from this dependency. Creating a system that pays its own way is an essential foundation for effective competition.'[70] Indeed this is the principle the Chinese government has nominally been promoting since the mid-1980s. Yet, creating a market for electricity has been difficult because of the current fragmented regulatory system. After all, 'political rules in place lead to economic rules.'[71]

Notes

1. Price, 1990, p.153.
2. North, 1990, p.45.
3. G. Fiancette and P. Penz, 'The Problems of Financing a Nuclear Program in Developing Countries', in IAEA, *Nuclear Power in Developing Countries: Its Potential Role and Strategies for Its Deployment*, Vienna: IAEA, 2000, p.153.
4. IAEA, 'International Status and Prospects of Nuclear Power', Vienna: IAEA, 2008, pp.29–30.
5. Luo Anren, 'An Economic Comparison between Nuclear and Coal Power Plants in Southeast China', JPRS-CEA-84–070, p.143.
6. Chang Jiming, Zhao Zhenggi and Chen Fumin, 'Nuclear Power Touted as Safe, Clean, Cheap Energy Source', *Zhejiang Daily* (31 August 1982), p.4 in 'China Report: Economic Affairs', JPRS 82976, 1 March 1984, p.108. Also see Luo Anren, 'Economic Advantages of Nuclear vs. Coal-Fired Power Plants Weighted', *Nuclear Power Engineering*, 4:6, 1983, pp.43–50 in 'China Report: Economic Affairs', JPRS-CEA-84–070, 24 August 1984, pp.144–157; 'Nuclear Power Plant to be Located at Hangzhou Bay', Xinhua News domestic service, 10 November 1982, in 'China Report: Economic Affairs', JPRS 82512, 21 December 1982, p.105; The Nuclear Energy Investigation and Research Group of the State Scientific and Technological Commission, 'A Proposal to Construct Nuclear Power Plants is Southeast China', *Beijing Daily*, 1 October 1980, p.4.
7. Luo Anren, 'Economic Advantages of Nuclear vs. Coal-Fired Power Plants Weighted', *Nuclear Power Engineering*, 4:6, 1983, p.145.
8. Yan Haiqun, 'On the Decline of the World Nuclear Energy Industry', *Beijing Shijie Jingji*, 9–10, 1984, pp.35–39 in 'China Report: Economic Affairs', JPRS-CEA-85–010, 29 January 1985, pp.14–22.
9. 李鹏, 2004.
10. Zhou Xiaoqian, 'Problems in Long-Term Planning of Power Industry to Year 2020', *Energy of China* (in Chinese), 3:25, 25 March 1993, pp.1–5, also in JPRS-CEN-93–007, 27 July 1993, pp.9–14.
11. IEA, 2006a, p.343.
12. IEA, 2009, p.160.
13. Joskow and Parsons, 2009, p.53.

14. World Nuclear Association, 'The Economics of Nuclear Power', information papers, November 2008, pp.8–9.
15. Ibid, p.6.
16. Ibid, p.2.
17. IEA, 2009, p.160.
18. IAEA, 'Nuclear Technology Review 2009', Vienna: IAEA, 2009, p.12.
19. NEA, 2008, pp.175–77.
20. NEA. 2000.
21. 肖建新 高世宪 韩文科, 我国核电实现跨越式发展的优势。挑战与建议, 中国能源, (Xiao Jianxin, Gao Shixian and Han Wenke, 'The Rapid Development of China's Nuclear Program: Challenges and Recommendations', *China Energy*), no.3, 2009.
22. IEA, 2007, p.358.
23. Berg and Tschirhart, 1988, p.3.
24. IAEA, *Financing of New Nuclear Power Plants*, Vienna: IAEA, 2008, p.1.
25. Newbery, 1999; Hunt, 2002; Xu, 2002.
26. Huang, 1996, p.63.
27. Smil, 1988, p.65.
28. Luo Anren, 'An Economic Comparison between Nuclear and Coal Power Plants in Southeast China', JPRS-CEA-84–070, p.144.
29. Yu Qiuli, 'Arrangements for the 1979 National Economic Plan', *China Report*, 15:6, 1979, 111–21.
30. Robert F. Dernberger, 'The Chinese Search for the Path of Self-Sustained Growth in the 1980s', in *China Under the Four Modernisations*, US Congress, Joint Economic Committee, 13 August 1982, p.40.
31. World Bank, 'The Private Sector and Power Generation in China', World Bank Discussion Paper No. 406, February 2000, pp.38–39.
32. Robert Delfs, 'The Balance of Power', *Far Eastern Economic Review*, 19 May 1983, p.80.
33. Andrew Fisher, 'China Provides the Power to Keep GEC Turbine Turning', *Financial Times*, 10 January 1986, p.6. Also see Andrew Fisher and David Marsh, 'GEC Set to Win 250 M Pounds Chinese N-Plant Order', *Financial Times*, 3 January 1986, p.1.
34. Louis do Rosario, 'Peking Gets its Way', *Far Eastern Economic Review*, 23 January 1986, p.48.
35. World Bank, *The Private Sector and Power Generation in China*, Washington, DC: The World Bank 2000, p.50.
36. During the 1980s, the central government and state-owned enterprises invested a total of 440 billion yuan in the energy industry, including mining, refining, transportation, and electric power. Electricity generating equipment accounted for 155 billion yuan of this amount. See Murray, et al., 'Foreign Firms in the Chinese Power Sector,' p. 641.
37. Speech of the vice-minister of the MEP on August 1, 1995 in the Planning Division of the MEP, Reform, Opening up and Planning in the Electric Power Industry (Beijing: China Electric Power Industry, 1996), p.143 (in Chinese); Fiona E. Murray, et al., 'Foreign Firms in the Chinese Power Sector: Economic and Environmental Impacts', in Michael B. McElroy, et al., eds., *Energizing China*, Boston: Harvard University Committee on Environment, 1998.
38. Quoted in Theresa Tan, 'Daya Bay: China's New Powerhouse', *Straits Times*, 6 March 1994.

39. Anonymous, 'China: Power', *Institutional Investor* 27:10, October 1993, 11. Yang Suping, 'Will There Be a Power Shortage in Guangdong'? *China Electric Power*, 555, June 2000, p.39.
40. David Schneider, et al., 'Power Plays', *China Business Review* 20:6, November/ December 1993, p.22.
41. 肖建新 高世先 韩文科, '我国核电实现跨越式发展的优势,挑战与建议', 中国能源, (Xiao Jianxin, Gao Shixian and Han Wenke, 'The Rapid Development of China's Nuclear Program: Challenges and Recommendations', *China Energy*), no.3, 2009.
42. Quoted from Ming Yang and Xin Yu, 'China's Power Management', *Energy Policy*, 24:8, 1996, p.747.
43. Paul S. Triolo, 'The Shandong Experiments', *China Business Review*, 23:5, September/October 1996, p.12.
44. SERC, 电力监管年度报告, 2008, (SERC, *Annual Report of Electricity Regulation*), April 2009.
45. Xu, 2002; Andrews-Speed, 2003; Hunt, 2002.
46. IEA, *China's Power Sector Reforms: Where to Next?* Paris: OECD, 2006, p.77.
47. Anderson, 2007, p.35.
48. 'Challenges and opportunities for US-China Cooperation on Climate Change', hearing before the Committee on Foreign Relations, US Senate, 4 June 2009.
49. 于得义, 华能集团核电战略的问题分析纪建议, 对外经济贸易大学, January 2007, p.21. (Yu Deyi, 'An Analysis of Huaneng Nuclear Strategy and Recommendations', Master Thesis of The University of International Business and Economics, 2007).
50. Kang Rixin was charged for taking bribes and corruption. The bond issuing was only a catalyst for his fall because after the global financial crisis swept the world, CNNC had heavy losses over its bonds in late 2008 and early 2009. Kang made a decision to use some of the money raised on the bond money to reinvest in stock markets to cover the losses, but ended with more losses. Kang was suspended when the National Audit Office investigated the stock market activities of CNNC. His corruption case came afterwards that led to his arrest in late 2009.
51. MIT, 'The Future of Nuclear Power', 2003, p.37.
52. NEA, *Nuclear Energy Outlook 2008*. Paris: OECD, 2008, p.203.
53. Patterson, 2007, p.125.
54. Liang and Goel, 1997, p.343.
55. Fridley, 1992, p.513, 518.
56. Ibid.
57. In December 1992, the State Council approved the collection of Y0.003/kwh for the construction of the Three Gorges project, proposed jointly by the Ministry of Finance, the SPC, the Ministry of Energy and the State Pricing Bureau. See the document issued by the State Council on June 14, 1993, at the legal document section of the SPCC's homepage, http://chinapw.cep.gov.cn. The provincial government was supposed to collect the two-cent construction fee. One cent would be used by provincial and local government for building power generation plants and the other cent was supposed to be used for the construction of transmission networks, which were in the hands of the MEP or provincial governments.
58. Ministry of Electric Power, *Electric Power Industry in China*, Beijing: China Electric Power Publishing, 1996.
59. Shi Jingping, 'Two Reforms and One Unification', *China Electric Power News*, 14 December 2000, p.4.
60. IEA, *China's Power Sector Reforms: where to next?* Paris: OECD, 2006, p.53.

61. 刘树杰, 杨娟, 陈扬, 彭苏颖, '核电价格形成机制研究', 中国物价, (Liu Shujie, Yang Jun, Peng Suying, 'An Analysis of Nuclear Power Pricing', *Chian Price*), October 2006, pp.18–22.
62. SERC, Ministry of Finance of China and World Bank, 'Study of Capacity Building of the Electricity Regulatory Agency', World Bank, working paper 45302, 21 August 2007, p.237.
63. NEA, *Nuclear Energy Outlook 2008*, Paris: OECD 2008, p.200.
64. SERC, 电力监管年度报告, 2008, (SERC, 'Annual Report of Electricity Regulation 2008'), April 2009, p.33.
65. Blackman and Wu, 'Foreign Direct Investment in China's Power Sector', p.704; Fridley, 'China's Energy Outlook', p.513.
66. Stephen Dow and Philip Andrews-Speed, 'Considerations for Foreign Investors in China's Electricity Sector', *Oil and Gas Law and taxation Review*, 11:8, 1998, p.311.
67. IEA, *China's Power Sector Reforms: Where to Next?* Paris: OECD, 2006, p.93.
68. 温鸿钧, '由核电与煤电外部成本比较看核电价格之优势', 中国核工业 (Wen Hongjun, 'A Comparative Study of Power Tariffs from Nuclear and Coal-Fired Stations', *China Nuclear Industry*), 4:60, 2005, pp.18–21.
69. NEA, *Nuclear Energy Outlook 2008*, Paris: OECD, 2008, p.176.
70. IEA, *China's Power Sector Reforms: Where to Next?* Paris: OECD, 2006, p.21.
71. North, 1990, p.48.

6
Technology Adoption or Technology Innovation

Nuclear energy development involves certain degrees of technology transfer, either through 'private-sector arrangements such as foreign direct investment (FDI), licensing, and joint ventures, or bilateral or multilateral technology agreements among governments.'[1] Some technology transfer leads to the adoption, innovation and development of indigenous technology. Technology imports, however, may also impede local innovation and development. Technology transfer and its standardisation and location can significantly bring down the cost and make it easier to regulate and therefore improve safety record.

To foster the development and diffusion of new technologies and the growth of new industrial capacities, the recipient country needs to have an enabling environment where the government and its policies can create 'greenhouses' that provide 'space for local entrepreneurs to experiment protected from transnational competition.'[2] At the core of this enabling environment is a set of coherent and strategic policies agreed upon by all government agencies.

What has been absent in China's nuclear energy development is any set of policies that can provide consistent guidelines for technology imports, standardisation and localisation. This is the result of (a) ideological debates between those who preferred the old self-reliance policy and those who emphasised the advantages of the market; (b) bureaucratic bickering; (c) rivalries within the nuclear industry and (d) competition for resources between those seeking 'quick returns' from technology imports and those wishing to develop domestic technological capacity. These problems have been compounded by the government's efforts to prevent the monopolistic control of the two players in the nuclear industry.

One of the early concerns about the transfer of nuclear technology to China for its energy programme was that 'China obviously ha[d] very good bomb designers; therefore, unlike practically every other developing country, it could make reliable, high yield weapons (at least in the kiloton equivalent range) from reactor grade plutonium.'[3] In retrospect, this assessment might have overestimated China's capacity in absorbing and upgrading the imported technology not so much because of the technical capacity of individual researchers but because of the erratic policies that made difficult to secure sufficient financial, human and material resources for the basic research and development.

Nuclear technology development in China is in great contrast to that in South Korea where a variety of nuclear reactors from France, Canada and the US was imported. South Korea, however, has successfully absorbed, adapted, upgraded these technologies and developed its own brand of large PWR (APR1400). The success of the nuclear industry in South Korea was possible because of a shared belief between the government and the industry that if the country wished to leapfrog towards modernisation, it should develop a nuclear industry, which, with its industry-wide technology spill-over effects, would enhance the productivity of capital, labour and other factors of production in the economy as a whole. Nuclear energy never meant reactor technology alone, but also its related fields – machinery, electrical equipment, basic design and architecture.[4] With the shared belief, the 'efficacious state [that] combined a well-developed, bureaucratic internal organisation with dense public-private ties'[5] was able to provide policy stability and specific financial incentives to foster and support large, powerful firms in their efforts of adopting and absorbing imported technologies.

China's failure in nuclear technology development also confirms how important consistent policies and enabling measures were. It had developed its nuclear weapon capability not because the country had resources to spare, but rather because of its concerted efforts to put resources together to support the nuclear programme. Once China started its civilian nuclear development, the consensus collapsed, not only on whether the country should launch a nuclear energy programme, but also on how to develop the industry.

The debate whether the country should put its resources into developing indigenous technology or importing the most advanced technology has divided policy makers and the nuclear community. It continues between those who prefer a 'one-step' process – introducing the most advanced types of nuclear reactor and standardising it in order to have the fastest development, and those who argue for a 'two-step' process – introducing a small number of advanced reactors and absorbing the technology, upgrading its own type, and then standardising it. In the process, China has built nuclear reactors from France, Canada, Russia and the US as well as its own, with key components supplied from more than a dozen countries. Domestic industries complain that they have been squeezed out by imports and research institutes complain that they are not given the opportunity to apply their innovation to nuclear projects.

Given that standardisation and localisation of technology is the main way to reduce costs and ensure the safe development of nuclear energy, the Chinese government in the Medium- and Long-term Nuclear Energy Development Plan (2005–20) stipulates that China will follow the path of 'introduction, digestion, absorption and re-innovation' in its nuclear technology development. This chapter will explore what China has been doing in terms of nuclear technology development, its successes and failures in technology imports and adoption, the forces behind the debate and the competition for different paths of technology development. It will also examine what China has been doing in meeting one of the most pressing challenges in its nuclear expansion – inadequate human capital.

Agreement on disagreements

One of the first priorities in the reform agenda in late 1970s was to 'restructure the research sector and to modernise indigenous scientific and technological capabilities.'[6] Energy, including nuclear power, was one of the 'new priorities for technology development' in the initiative. Introducing technology from abroad was advocated by those who charted the reform: 'We should *introduce* selected advanced technologies that play a key and pace-setting role in line with the needs for modernising our country,' announced China's Minister of the State Science and Technology Commission (SSTC).[7]

This idea of 'introducing' technologies was interpreted quite differently. There were, by and large, three positions in terms of nuclear energy development. The first position held that nuclear energy development should rely only on self-designed, self-manufactured and self-managed and operated projects. The second asked, as other countries had already developed mature technology for safe and reliable NPPs, why reinvent the wheel? Importing turnkey NPPs would allow China to build a nuclear programme quickly and in the process learn how to make its own NPPs. The third position argued that China had fallen behind in nuclear technology and it was necessary to import the most advanced technology, but the focus should be on how to develop its own capacity for designing, manufacturing, constructing and managing NPPs. This debate carries on today.

The argument based on the strict interpretation of self-reliance lost its appeal quickly as the first NPP in China – Qinshan I – progressed. Those who had been working on HWRs for the submarine programme insisted that China should build its nuclear energy programme based on HWR technology it had already developed and limit imports to the minimum to avoid becoming dependent on foreign technology. 'We must rely mainly on our own efforts while making foreign assistance subsidiary,'[8] they argued. They were criticised for continuing 'to look back fondly on its previous successes in science and technology, primarily the development of the country's nuclear weapons and intercontinental ballistic missiles.'[9] Their opponents argued that PWRs were used in most countries and should be the technology for China. To do so might require initial imports of Western technologies and substantial international assistance and cooperation, but Chinese scientists could build their own brand name of technologies through imports.

Qinshan I was built on this principle: it was a PWR with a 300MW capacity, designed by the Shanghai Nuclear Engineering Research and Design Institute (SNEDRI) and constructed by a Chinese team, but many important components were imported: the pressure vessel was manufactured by Mitsubishi in Japan to Chinese designers' specifications from a Westinghouse design; the polar crane and reactor coolant pumps were imported from West Germany, and tubing for the turbine generators came from Sweden.

The proponents of this revised version of the 'self-reliance' approach argued that, given sufficient time and resources, Chinese scientists would be able to develop their own brand name of designs and to manufacture, construct and operate NPPs as other countries did without further resorting to foreign technology. Their insistence on using domestic designs and limiting imports led to an eruption of disagreement with MWREP in the process of constructing Qinshan I. MWREP was granted overall jurisdiction for constructing the NPP while MNI was delegated primary responsibility to build nuclear islands (the nuclear reactor and primary cooling system). MNI insisted on the importance of indigenous technology even though it would take longer and cost more. While MWREP preferred a quick development to alleviate the pressures of power shortages, MNI wanted to see the industry thrive. Their disagreement enhanced the unwillingness of MWREP to take nuclear as its solution to power shortages. It is important to note that competition among bureaucracies, nuclear industries and other related sectors over reactor designs was common in all countries with nuclear energy programmes. It was novel, however, in the early 1980s in China where disagreements among government bureaucracies were seldom made open.[10]

The opponents to the self-reliance approach argued that even if Chinese scientists were able to design, manufacture, build and operate their 300MW and 600MW models, if no one else adopted them, the technology would not have the economies of scale or safety record to be able to compete with Western technologies. They cited nuclear development in Britain as an example. Britain started researching and manufacturing graphite moderated gas-cooled reactors (Magnox). But, when Britain decided not to be part of the Euratom and the Germans first ordered a boiling water reactor (BWR) from the American General Electric, it became clear that the Magnox could not be used as the base model for nuclear development. France decided to stay away from the technology too and turned to light water reactors. 'The decision stemmed from a different consideration: a realisation that unless a switch was made, France would be cut off from the benefit of the operating experience which was being accumulated around the world, and in all probability from world markets for reactor exports.'[11] Framatome took a Westinghouse licence as a practical means of obtaining the new technology. By the time the oil crisis came, the French parliament authorised a sharply increased annual rate of ordering of five or six nuclear stations a year because 'Framatome's PWR proved cheaper.'[12]

Those from Guangdong passionately argued for importing a turnkey NPP for the Daya Bay project because they wanted to have it built with the most advanced technology. A turnkey project would bring power on stream sooner than would a contract with several suppliers for various components. It would take two decades, some argued, for domestic technology even to get close to what Western nuclear vendors could offer. This position to a large extent was shaped by the situation in Guangdong, which was the first province that pioneered opening up to the outside world and introducing foreign investment. A turnkey project would not only bring in the hardware the country needed but also improve its 'software' as its people would be trained by the international vendors and as local people would watch the Western technicians build and operate the nuclear power plant.

In addition to a nuclear island from Framatome and a conventional island from the Franco-British joint venture GEC Alsthom, the contract for the Daya Bay project put in place a comprehensive training programme for Chinese technicians and professionals. A French civil engineering consortium was in charge of construction and EDF managed the operation and provided technical assistance in project management and the coordination of start-up. All 800 workers employed at the power plant underwent some form of training, either in China or in France. Three hundred people (half fresh from universities and half from operating facilities and research institutes) were recruited and trained as shift supervisors and deputy managers by experts from EDF in China. Twenty of them were sent for 18-months training at EDF nuclear plants in Blayais, Cattenom and Chinon. Another 130 or so were sent to France and Britain for 6-months training in PWR principles and on partial and full-scope simulators.

Another 350 candidates were recruited from universities and technical colleges and given specific training in China, France and Britain in such fields as fuel handling and sludge lancing. The rest were trained mainly in China by French and British experts as field operators and support staff. France's Institut de Protection et de Sureté Nucléaire (IPSN) cooperated with China's National Nuclear Safety Administration (NNSA) mainly for safety analysis of the two units. Some 20 Chinese safety experts received several months training at IPSN's headquarters to work on the plant's preliminary safety reports, design document evaluation, prepare final safety reports, and assess a start-up test programme and their results.[13] By the time the Daya Bay NPPs went into commercial operation, CGNPC had built a team of engineers, control room operators and management staff who had gathered a rigorous five-year training programme supervised by people from EDF. They were ready to take on a next project.

The development strategy preferred by people in Guangdong was shaped by their reform experience and the experience in nuclear development in France and Japan, both of which had imported the Westinghouse PWR and then built their own nuclear industry. They were particularly impressed by the nuclear development experience in South Korea that shows the imported technology from turnkey projects could be standardised and localised through a learning-by-doing process, and then develop its own unique model.[14] These are the experiences people in Guangdong decided to emulate. CGNPC would import another turnkey project, increase the domestically manufactured components from zero to 30%, and standard-ise the design and manufacturing to fit local conditions. Eventually, they would be able to localise the technology for nuclear expansion in China.

The third view was held largely by the nuclear science community, which rejected the strict 'self-reliance' approach and opposed the turnkey approach too. This position was discussed by Wang Ganchang, one of China's fore-most nuclear scientists and president of the Chinese Nuclear Society in the early 1980s:

> We might seek technical help from a friendly country well-advanced in nuclear power and undertake with her an all-out cooperative program, like that between Brazil and the Federal Republic of Germany, so as to realise technical transfer in the shortest time, leading to a capacity of designing, building, and operating a commercial nuclear power plant by ourselves in the 1990s. As an alternative, we might also choose to rely mainly on self-reliance with a limited amount of technical help from abroad. In this alternative, demonstration nuclear power plants of smaller capacity are to be built and operated as a first step before full-sized commercial plants are constructed.[15]

In discussing the two alternatives, Wang made it clear that the second alternative would significantly delay the start-up of China's nuclear energy development. Delay in the nuclear programme would 'not only be detrimen-tal to the development of nuclear industry but also bring about waste and loss among the nuclear power science and technology forces,' many in the nuclear community argued. 'Such losses would be irreparable.'[16] According to Wang, China should adopt the first strategy – negotiating with other countries for cooperation in building NPPs. Importing foreign technology was necessary, but it should include heavy components of technology trans-fer as a condition. China then could capitalise on the existing research and human capacity to manage a process of innovation and improvement.

They opposed turnkey projects because, as they argued, 'when you buy a bag of rice from a shop, you do not buy the ability to grow rice'. If China wanted to have a nuclear industry, it would have to develop its own capac-ity to design, manufacture, operate and manage NPPs. Technology imports therefore must be conditional on technology transfer, while turnkey projects would minimise technology transfer. 'Even under the current reform and opening up policy, China cannot depend on the import of nuclear power stations,' Ouyang Yu, the chief engineer and designer of the Qinshan NPP, said. 'It should master the complete technology needed in building and running a nuclear power plant, including the research and design and the skills to manufacture, construct, manage and operate a plant.'[17] Self-reliance remained their basic idea, but self-reliance did not or should not preclude foreign cooperation. They followed Mao's interpretation of self-reliance – 'keep the initiative in your own hands.'[18]

This became the dominant position among people in various research institutes under MNI and later CNNC. It won over many in the first group, especially after the performance of these domestic components for the Qinshan project adversely affected the viability of the plant and many components had to be replaced. The indigenously manufactured components (mainly for the non-nuclear balance of plant generating equipment) were for the most part 'off the shelf' and therefore not designed specifically for the Qinshan facility. This experience with domestic components prompted the NNSA to initiate a vendor quality assurance programme. For the Chinese designers, engineers and constructors who had not undertaken such a project before, this project was necessary to shorten the 'learning experience' in building nuclear power plants.[19] Many others, however, recommended the greater use of imports to enable Chinese scientists to adopt and absorb the advanced technology. This position was shaped as much by the history of China's nuclear development as by the organisational culture of the nuclear sector, which had emerged from a military programme shrouded in secrecy and was much more closed than Guangdong province.

The two positions were apparent along regional lines too: those in Guangdong pursued the strategy of importing foreign equipment and technology to minimise the time needed to master the technology while those at CNNC preferred to develop indigenous capabilities by incorporating the best technology available on the world market. The two groups agreed to disagree in the 1980–90s: after Qinshan I, CNNC would start another project with predominantly domestic design and manufacturing, the CNP series, while CGNPC would import another turnkey project, the CPR series.

In 1987, when the State Council approved in principle a new project at the Qinshan site, it made it clear that the new project would be designed, manufactured and constructed by the Chinese, scaling up the technology to 600MW, with some international cooperation (CNP600). Like Qinshan I, this project was designed by Chinese engineers from several research institutes attached to the then MNI, the MEP, particularly the Beijing Institute of Nuclear Engineering, and the Shanghai Boiler Works Corporation, the Harbin Turbine Works Corporation and others. They undertook the design of 47 out of the 55 key components of the reactor, accounting for 55% of the total components.

Unlike Qinshan I, which was extrapolated from a Chinese nuclear submarine reactor design, Qinshan II was the scaling-up reactor of the model used at Qinshan I and adopted the PWR used at Daya Bay as a reference design. Engineers then modified them to suit differences in site conditions, capacity, and grid requirements and with a desire to improve on the French design. Even though this meant only about 40% of the reactor's island, 60% of the conventional island and 76% of turbines were designed and constructed domestically, the project was proudly announced by the Chinese as 'a new step toward self-reliance' and 'one of the major engineering projects of the Chinese state.'[20]

To make this upgraded model of reactor possible, the Chinese needed more experience. The opportunity came when Pakistan approached China to build a replica of the 300MW PWR at Qinshan. China and Pakistan signed a preliminary contract on 31 December 1991, just 15 days after Qinshan I was connected to the grid. With the agreement, China provided and built a nuclear power plant with 300MW capacity in Chashma, supplied nuclear fuel to run it, and transferred nuclear technology to Pakistan. China also provided funds for foreign exchange needs while Pakistan paid for the local expenses. In March 1992, China acceded to the Non-Proliferation Treaty, partly to place this project under the IAEA safeguard provisions. On 24 February 1993, China, Pakistan and IAEA signed the Safeguard Agreement for Chashma. The project went ahead.

This contract to build a NPP in Pakistan was significant for the Chinese nuclear industry at the time because: (a) it was the first time ever that China had exported nuclear technology and equipment; (b) it was the largest order the nuclear industry had received in China and (c) it received an order at a time when the industry enjoyed the least growth among all industries. There was a hope that by building another 300MW PWR, the Chinese would make this model of reactor with both 300MW and 600MW capacity standardised. Indeed, in the early 1990s, many in the sector believed that a small to medium-size PWR would be suitable for China because the country had limited financial resources to support large expensive units and their smaller size also meant that there would be less pressure on the grids. They would also put less pressure on human capital too.[21]

Ironically, nearly 20 years later, as many Chinese scientists and policy makers have moved on and endorsed large-sized and more advanced models of reactor (AP1000 and EPR), small and medium-sized reactors with the equivalent electric power of less than 700MW provide an attractive and affordable nuclear power option for many developing countries with small electric grids, insufficient infrastructure and limited invest capability. IAEA has listed Chinese CNP 300MW and 600MW as one of the options even though China has decided not to build them anymore. The Chinese government and its nuclear industry have closed the door on their future opportunities.

CGNPC pursued its own alternative strategy, importing French reactors while increasing the share of domestic components. After Daya Bay went into commercial operation, CGNPC proposed another project in Guangdong (Lingao, 5 km away from Daya Bay) with turnkey reactors but with an increasing share of domestically manufactured components, from 1% in Daya Bay to 30% in Lingao. Like Daya Bay, the nuclear reactor for Lingao was designed, manufactured and installed by Framatome and non-nuclear facilities by GEC Alsthom. Unlike Daya Bay, construction and assembly work at Lingao was undertaken by Chinese enterprises under Framatome's and GEC Alsthom's supervision and technical assistance. The same people of CGNPC who managed Daya Bay would manage Lingao.

The unforeseeable political developments in 1989 and renewed isolation of China from the international community changed the context in which nuclear policies were made. China signed on with Russia to import VVER in the Tianwan project and with Canada to import CANDU HWRs for Qinshan III, primarily for political and diplomatic reasons. Opposition to introducing Russian VVER reactors came from several directions: some were concerned about the safety record of the reactor design, which had been used in the Chernobyl NPPs. They did not think China should take the risk only four or five years after the meltdown. Others were concerned that, given Russia had not built any new NPPs for more than two decades and the whole country was in political chaos, it was questionable whether the country could deliver safe nuclear reactors on time. These turned out to be real problems once the project started.

'The Russians might have had some 50 years of experience in the nuclear industry, but they had not built a nuclear power plant for over 20 years,' recalled some Chinese officials. 'In addition, RRV had incorporated some new technologies that became real challenges for the Russians. On many occasions, we had to come in and help because they could not solve the problems.'[22] The chairman of the Tianwan project board later admitted, 'This is not a commercial project, but a *political* task.'[23] Introducing a Russian reactor did not help China standardise the technology at all. It created many problems, among which was the intensified debate on technology selection: China had already adopted PWRs from Framatome, why did the government decide to waste money on a problematic technology instead of allocating resources to support domestic institutions in upgrading its own PWR used in Qinshan I?

This was also the argument used for the Qinshan III project, which used CANDU reactors. The difference was that while there was no support in the nuclear industry for importing the Russian model, a group of experts from China's weapons programme wanted to continue the HWR technology. They had been marginalised in nuclear energy development, yet after 1989 their preference for HWRs coincided with those who wanted to resume China's relationship with the West and end the new isolation. Others welcomed the importation of the CANDU reactor because they saw nuclear fuel supply as a major obstacle for China's nuclear development. CANDU reactors do not require enriched uranium and CANDU components were easier to fabricate than major PWR components and the fuel was simpler because on-load refuelling provided more fuel cycle flexibility. This was also the time when many reformers in China were attracted to the development models of Japan and South Korea, both of which had imported CANDU reactors.

For many in the industry, Tianwan (with Russia) and Qinshan III (with Canada) were the two worst cases in technology development. Neither contract included provisions for technology transfer; the Chinese counterparts, indeed, had no intention of even learning the technology. Import for the sake of import, argued the opposition, was opening the door for foreign companies while killing China's own industry.[24] Yet few people could argue against their political importance.

In sum, in the first two decades of its nuclear energy development, the debate on how to develop the Chinese nuclear industry and the best route by which to do so was largely conducted among those in the industry and within a limited circle of policy makers. Decisions were shaped by the general political situation in the country, while the emphasis on developing self-capacity was upheld. CNNC and CGNPC did not agree on what indigenous technology meant – building from scratch or on imported technology. They did agree that the industry would have a future only when they were able to make the technology standardised and localised. Nonetheless, they could not agree on the best way to achieve their aims. Those at CGNPC believed in the path taken by South Korea and Japan, while those at CNNC

Reactor generations

			Generation IV
		Generation III+	Revolutionary designs
	Generation III	Evolutionary designs	
Generation II	Advanced LWRs		
Generation I	Commercial power reactors			
Early prototype reactors			
Shipping port Dresden Magnox	PWRs BWRs CANDU	CANDU6 System 80+ AP600	ABWR ACR1000 AP1000 EPR ESBWR	Enhanced safety. Minimisation of waste and better use of natural resources. More economical. Improved proliferation resistance and physical protection.

Gen I	Gen II	Gen III	Gen III+	Gen IV

1950	1960	1970	1980	1990	2000	2010	2020	2030	

Source: NEA, *Nuclear Energy Outlook 2008*, Paris: OECD 2008, p.373.

wanted to have a Chinese brand rather than making something that had been designed by others. By the end of 1990s, both CNNC and CGNPC had decided to build their own upgraded Generation II reactors: CNNC would build on its CNP300 and CNP600 while CGNPC would start working on CPR1000, based on the French M310 model.

Continuing debates in the 2000s

The agreement to disagree on the two paths of technology development was threatened in the 2000s when the central government showed growing interest in nuclear energy development and different government agencies began to push forward their own views in the decision-making process.

At the end of the 1990s, there was an outcry in the nuclear community that China had adopted too many types of reactors that made standardisation and regulation so difficult. 'The proliferation of plant types increases the probability of error,' advised an American consultant. 'Without standardisation, maintaining safety can become extremely complicated, and plant operation and maintenance costs can become excessive, as such design requires regulators, regulations, and maintenance and training procedures tailored to that plant's characteristics.'[25]

A joint study published in 1999 represented the beginning of an era when think tanks can make a significant contribution to policy making. A team of experts, led by the Development Research Centre of the State Council, discussed the issues of technology selection. It pointed out that the 'two-leg' policy was an official policy on nuclear energy development: the policy of 'self-design, self-manufacture, self-construct and self-operate' (自主设计, 自主制造, 自主建设, 自主运营) and that of 'self-reliance as priority, international cooperation, introducing advanced technology and promoting indigenous technology'. To simplify, it is a combination of self-reliance and the importation of advanced technology. This might be a correct strategy, but in practice, it failed completely.

The report highlighted three major problems of technology adoption and adaptation in China: (a) the eleven nuclear reactors in China came in five models and from four countries; (b) financing was in favour of imported technology at the expense of domestic technology development and (c) each NPP project had its own plan in technology adoption and application and there was little coordination. Introducing varieties of technology not only raised costs of NPP development but also created many difficulties in construction, operation, maintenance, human resources and regulation. More importantly, it impeded the development of indigenous technology.

The report acknowledged the complicated reasons for this development, which included lack of a national strategy, financial constraints and contradictions among several sectors in the economy.[26] The key to all the problems was the lack of a single institution that had the expertise and the authority to make decisions on technology selection. Many organisations were involved in making decisions on nuclear energy: SPC, SETC, CONSTIND, CNNC, CGNPC, the Ministry of Science and Technology, the Ministry of Finance and many others, while each had its own interests to pursue and its own limitations in making decisions.

In January 2000, the central government called a meeting of five leading Chinese nuclear organisations (CNNC, the Ministry of Machinery, the Beijing Institute of Nuclear Engineering, also known as the Second Engineering Institute of Nuclear Energy, the SPCC and CGNPC) 'to pool their resources and collaboration on the development of a standardised commercial power reactor which might be ordered sometime during the 10th FYP (2001–05).'[27] At the meeting, participants reached an understanding that China would not order turnkey nuclear power reactors anymore; instead it would work on the standardisation and localisation of the two main models used in Qinshan I and II (CNP300 and CNP600) and in Daya Bay and Lingao (CPR1000 based on the French M310). With this understanding, CNNC would work on the extension at the Qinshan site and build a new model of CNP1000, while CGNPC would start work on another two units at the Lingao site by adapting and upgrading the technology imported from Framatome and increasing the portion of domestic components.[28]

In 1999, CNNC had been notified that the site for its upgraded CNP1000 would be moved from Qinshan to Sanmen in Zhejiang province. In the following five years, CNNC invested 90 million yuan and also borrowed a total of 500 million yuan from two of the four state-owned banks to prepare for the project at Sanmen. It gathered a group of some 70 experts to develop and design its CNP1000. CNNC also reached an agreement with CGNPC in 2002 to cooperate on upgrading CNP600 to CNP1000 and the M310 to CPR1000. This cooperation might have been pushed by the central government to rationalise R&D in nuclear technology. It made sense for the nuclear industry for both political and technical reasons. Politically, Premier Zhu Rongji was not about to approve any nuclear projects before leaving office in mid-2003. While still in office, Zhu charted the economic course to shunt hydroelectric power from central and western China and coal-fired electricity from Guizhou to Guangdong in order to spread economic development to the poor, land-locked regions. CNNC and CGNPC needed a combined voice to protect themselves.

While major efforts were made along the two lines for the extensions of Qinshan II and Lingao, in early 2002 the State Development and Planning Commission, the predecessor of NDRC, issued a document, No. 2866, stating that China's nuclear energy development would build on its indigenous CNP600 and develop CNP1000, while introducing the most advanced technology if necessary. Many in the industry interpreted the document as encouragement and support of the central government for domestically developed nuclear power reactors. Of course, this was not what CGNPC preferred. Neither CNNC nor CGNPC was ready to give up its development strategy and endorse the other as the base model for China's nuclear energy development.

Less than a year later, in April 2003, SDPC announced that China would call for international bidding for two nuclear power projects at Sanmen in Zhejiang and Yangjiang in Guangdong with two units each. Many assumed that the bidding would be for the most popular model of reactors of Generation II. SDPC emphasised that the objective for international bidding was to introduce technical and engineering expertise and reactor components on the principle of 'combining technology and trade, and market for technology' (技贸结合, 以市场换技术). The announcement caused uproar in the Chinese industry which had believed that the dual track strategy had been the accepted policy and that the country would not import any new models. The decision, however, was not a surprise in the international nuclear community because the word had already been out in 2002 that China would import foreign technology to fulfil its nuclear ambition set in the 10th FYP – to build 36GWe capacity by 2020 with 18GWe under construction.

To prepare for the bidding, CNNC and CGNPC brought in China Technology Import and Export Corporation (CTIEC). Before their work could go anywhere, there was a change of government. The team of Hu Jintao and Wen Jiabao replaced Jiang Zemin and Zhu Rongji and SDPC was renamed as NDRC. In March 2004, just days before US vice-president Cheney visited Beijing to discuss nuclear energy cooperation with the Politburo, NDRC announced that, instead of introducing advanced Generation II technology, China would call for international bidding for Generation III PWR technology.[29] This excluded the Canadian CANDU, which is a heavy water reactor. It also shut the door on CNNC and CGNC for the new projects because both CNP1000 and the CPR1000 were Generation II technology.

NDRC did not explain why it had changed the policy except that the country needed the most advanced technology for its nuclear development so that eventually China would become a leader in the field. Neither the public, media or those in the nuclear industry was satisfied with the explanation. Many speculations then emerged: some suspected this change had a lot to do with the change of government from the Jiang-Zhu team to the Hu-Wen team, which was quietly making many changes regarding energy policies. Some suspected that as the electricity shortage spread across the country, NDRC had decided that the quick expansion of nuclear generation capacity depended on the adoption of the most advanced technology (this explanation, of course, does not stand up since it would always take longer for a new model of reactors to be constructed and installed). Many were convinced that the decision was made because the Bush administration had signalled that it might actually decide to sell civil nuclear technology to China.

China had always wanted to introduce nuclear technology from the US and it had negotiated with every administration since Ronald Reagan's presidency, but the deals had not been given the green light by Congress. When there was a sign that the Bush administration might change its stand on the issue, China grabbed the opportunity. The nuclear industry in the US had lobbied fiercely for more than two decades to get into the Chinese market. Their representatives told Congress that if the ban of nuclear technology transfers to China was extended indefinitely, 'it could mark the beginning of the end for the US and other Western nuclear-plant builders whose markets had dried up at home.'[30] Between 1998 and 2000, 16 applications for licences were filed to the Nuclear Regulatory Commission (NRC) and the Department of Energy to export US nuclear technology to China.

At IAEA in Vienna, increasing diplomatic exchanges took place between the Americans and the Chinese, who assured the Americans that 'US nuclear technology would not be re-transferred by China to third parties without prior US consent.'[31] In 2002, when China announced that it would put two nuclear projects up for international bidding, the US embassy in Beijing reported back to Washington that it believed 'about 300 enterprises are engaging in the development and production of nuclear technology in China.'[32] For the Americans, this meant that billions of dollars were at stake. The Bush administration decided to help Westinghouse win the bid.

"The US government began working with the Chinese government to support the bid of a US manufacturing in 2004."[33] The Department of Energy had a cost-sharing agreement with Westinghouse for the AP1000 design, the completion of the NRC design certification and for the NRC licensing and construction of the first standard AP1000 nuclear plant design. The NRC fast-tracked the approval process for the AP1000 specifically so that Westinghouse could participate in the bidding for the Chinese contract. Just before the formal bid was filed, 'the Export-Import Bank approved a Preliminary Commitment for guaranteed and/or direct loans of up to US$5 billion to support Westinghouse.'[34] 'Secretary of State Condoleezza Rice and Secretary of Commerce Carlos Gutierrez contributed support to Westinghouse's bid,' too.[35] Vice-president Cheney visited China in 2004, in an orchestrated media campaign to condemn North Korea's nuclear programme, but he spent much of his time praising Westinghouse's new AP1000 design. By then, it was clear that the US Government linked the bid with the issue of China's trade surplus and its currency appreciation. 'Coincidentally', NDRC in 2004 revised its decision from Generation II to Generation III reactors.

In the 1990s, slow nuclear power development led to a significant consolidation of the nuclear power construction industry. General Electric joined forces with Hitachi, Framatome with Siemens, and Westinghouse with British Nuclear Fuels plc (BNFL), and later Toshiba. Russia's Atomstroy-export (ASE) and Canada's AECL remained in the field. This consolidation had two major consequences: a limited capacity to construct nuclear power stations and a focus on pressurised water reactors by only a few manufacturers joining forces in competing for the shrinking market. Given the size of the con-tracts and their importance in terms of each country's domestic economy, they all had strong backing from their respective governments to enter the potentially lucrative Chinese market.

The competition for Chinese contracts was well reported in China. Yet, unlike the debate over the site selection of NPPs, as discussed in Chapter 8, the debate on technology importation and adoption was conducted among the elite – a few top policy makers and those in the nuclear field. Articles and comments emerged in professional journals, and newspapers, the inter-net and magazines. For those who cared to seek it out, there was plenty of information about opposing views; the debate among elites was intense and both sides made their views clear in the public arena.

The issue, though, did not affect the general public and indeed the public rarely entered the debate. This development had two implications for policy making: one was that the debate was relatively self-controlled, not because the government or the party suppressed the argument, rather because those in the debate were important political players. Their fights were 'covert' and took place behind the scenes. This leads to the second implication: compro-mises were made with little public explanation. Indeed, it is not clear if the Politburo was even aware of the opposing views and/or how seriously they took the debates. The core issue for those involved in the industry was one of its survival, while the core issue for the top political leaders was the high diplomacy and power politics among major players in the world.

One argument provided to support the imports was that AP1000 and EPR were much better technologies than anything the Chinese had developed and would develop soon. Some bureaucrats in the nuclear sector publicly claimed that there was no way that their own model would ever be competitive with the Westinghouse AP1000. The opposition was mostly at the elite level in the nuclear industry too. They were eventually overwhelmed in policy terms but their arguments are important because they represent challenges that China faces in its nuclear expansion. Some of the arguments also highlight the core issues that apply to nuclear development throughout the world.

First, one question that was often asked by Chinese scholars and the media was: why were some government officials so eager to push forward more imports of nuclear reactors while the country had already demonstrated that it could build its own? Without naming individual officials, they asked why some government officials had become salesmen for foreign companies in China and what benefits they had received from these Western vendors for pushing through the deal.[36]

Nuclear projects are expensive and capital-intensive. When China plans to build two or three units a year for the next decade, this means 50–70 billion yuan are up for grabs. Who would benefit from this development – domestic or foreign players? When some government officials pushed for international bidding, did they do so because they believed it would serve the country's best interests or were they driven by their institutional interests or even for personal gain? These questions might sound unreasonable, because at the time there was no evidence to suggest that decisions were made with any bad intentions. Yet, corruption cases that were later discovered confirmed the suspicion (see later in this chapter).

Second, there was also a major argument about the future of the domestic nuclear industry. The nuclear industry is not only about reactors. Related industries would also be affected if China gave the whole project to foreign companies, as stated by the NDRC's decision. For example, for two units of a 1000MW nuclear power stations, there would need to be at least 30 000 valves, with diameters ranging between 0.5 and 440 millimetres, and pressure levels ranging between 150 pounds and 2500 pounds. The valves would be made from different materials, from plastic to stainless steel, and powered by either electricity or gas. Two per cent of the total cost of a nuclear power station is attributable to valves in all sizes and kinds and often 50% of the maintenance cost of a nuclear power station is on valves.[37]

Granting nuclear power contracts to foreign companies would endanger all the similar auxiliary industries. With this backdrop, the five major nuclear equipment makers in China – Shanghai Electrical Corp., Harbin Electrical Corp., China First Heavy Industries Corp., Orient Group and China Second Heavy Industries – launched their campaign to secure the government's support. They already had experience in producing turbine generators and other auxiliary facilities, a core part of the Chinese economy.[38] They asked, why had the government decided to deprive the Chinese companies of opportunities and give the jobs to foreign companies?

For those Chinese scientists who had been working on CNP1000 and CPR1000, the decision was hard to swallow, especially considering that all international bidders had active government support from their home countries. The official newspaper *China Daily* reported on 23 February 2005 that all bidders 'boasted firm support from their own governments' and the claim was supported by the Western media as well: there was 'very heavy-handed engagement' by supporting governments to try to ensure that China would favour their 'national champions.'[39]

Some in the Chinese nuclear industry argued, introducing so-called Generation III reactors would spell the end of the industry, which would never be independent in developing its own brand name of technology, in the same way as occurred in the automobile industry and the civilian aeroplane industry. In both fields, the supporters of foreign designs and technologies had won and the Chinese simply did not have the opportunity or support to develop their own industries. The fields were 'granted' to foreign companies, while Chinese companies simply became 'makers' of the Western products. Little technological capacity was developed in the process. In contrast, because the US had banned the export of military jets to China, the Chinese military and scientists had been able to develop a top quality jet comparable with products from other countries.

Third, there was an issue of safety and security. One major criticism of the decision was that NDRC placed the country's nuclear industry at *technical, financial* and *security* risks.[40] According to some Chinese nuclear scientists, neither AP1000 nor EPR was the IAEA-approved Generation III technology. As one Chinese scientist, who served as the deputy director-general of IAEA for more than 10 years (1992–2002), explained to some people from Westinghouse:

> We started the idea of CNP600 at the same time you started your AP600. This means that we entered the school at the same time. It was just you were rich and had the money to continue and we did not. Now you have graduated to the next level and have developed your AP1000. But, please don't treat me like an idiot. Your AP1000 can hardly be called as Generation III despite all the advantages in terms of safety. There are no fundamental changes in terms of technology from AP600.[41]

In addition, there were no precedents. Areva was granted a licence to build an EPR in Finland in December 2003, but even in its early stages, there were signs of problems with rising costs and delays. If EPR were so advanced and attractive, why didn't France have one in operation? No one country had built an AP1000 either. No American companies had been willing to adopt the technology and build one AP1000 in the US. Japan had looked into AP1000 but dropped it around the time when the Chinese government decided to pursue it. Why would China want to take the risk and become the guinea pig for an untested American product?

There would be no experience from which to draw or lessons from which to learn. Two key requirements of any nuclear reactors are their safety and their economy. AP1000 almost doubled the price of CNP600 used for Qinshan II. Moreover, the vast majority of reactors operating in the world are Generation II designs, which, after many years of tests and trials, have proved to be safe and economical. If many countries could have built their nuclear programme with the technology, China should have been able to do so as well. As for the safety concerns, China already had experience in building and operating Generation II reactors, but no one had experience in constructing and operating AP1000, as General III reactor.

Fourth, nationalism entered the debate. Media drew attention to the fact that Westinghouse was no longer an American company because it had been bought by the Japanese company Toshiba. If the bid was won by Westinghouse, it was argued, China would face political and security risks of subjecting its nuclear industry to the Japanese control. After Westinghouse put in its bid for the Sanmen and Yangjiang projects, Toshiba beat other well-known companies, such as General Electric and Mitsubishi Heavy Industries, in purchasing Westinghouse for US$5.4 billion in early 2006, almost tripling the price the UK-government-owned British Nuclear Fuels had paid when it bought Westinghouse in 1999.

The CEO of Toshiba made it clear that it had paid a hefty sum for the 100% stake in Westinghouse with every expectation that nuclear power generation would expand quickly, especially in China.[42] The fear was expressed by the media that giving Westinghouse the contract would mean that the Japanese would enter into the Chinese nuclear market and fulfil its ambition to control China's nuclear development. This issue became a hot political potato for the Chinese government when negotiations with Westinghouse were progressing. The government was not prepared to trigger the public protect by granting the deal to a Japanese company. To calm anxiety in China, Westinghouse officials announced in public that Toshiba 'would allow Westinghouse to operate as if it was an independent US company' and 'Westinghouse would not acquire a Japanese label in China.'[43]

Finally, the issue that some policy makers, especially at NDRC, had deliberately changed policies agreed upon a couple of years earlier arose: that is, China would build its model based on its indigenous technology, either CNP or CPR. While foreign firms were jockeying for influence in decisions on future contracts and new construction projects, 'a cat fight' broke out 'at the very top of the government' in China.[44] Among those involved in the disputes included the top managers of large state-owned corporations, particularly CNNC and CGNPC. As far as they and others in the industry were concerned, the central government had already decided in the late 1990s that the country would not import any more turnkey projects. Why did NDRC change the previous policy?

In 2002, when visiting Qinshan II, Hu Jintao, then vice-president of the country, stated that the 'nuclear energy industry is a strategic industry and China needs to develop its own technology for its expansion. No money can buy the core technology. Developing indigenous design and technology is the only way for nuclear expansion'. This speech was consistent with the spirit of a document issued early in the year by the NDRC. The 10th FYP (2001–05) also outlined the strategy for nuclear development of 'combining self-reliance with foreign partnership.'[45]

The official policy was to rely on indigenous technology. This was incorporated in the draft short- and medium-term plan for nuclear energy development, and NDRC was to carry it out. Statements by officials and details of the policies were quoted and requoted by the opponents of the deal in their interviews and by the media. Many people in the nuclear sector resented the enthusiasm shown by some government officials, especially those from NDRC, to push foreign technology down their throats and to change previously agreed policies. Meanwhile, to put more pressure on the industry, NDRC in 2006 announced that no nuclear power plants in interior provinces would be approved without using GIII reactors, AP1000 or EPR1000.

This bitter fight, no matter how discreet it was supposed to be, was widely reported in both China and the Western media. Even some officials at the US Department of Energy admitted that the debate over the introduction of AP1000 in China was political rather than technical.

In early 2006 while the international bidding was still under way, some officials from the NNSA explained the Chinese policy on nuclear technology development at an international conference in Moscow:

> In order to help increase the domestic capability to design, manufacture and build nuclear power plants independently, some advanced nuclear power technologies (generation 3, G3) will need to be introduced by international bidding. However, for meeting the enormously increased need for electricity in the near future, China will have to continue to build some generation 2+ (G2+) plants and base their design, subject to modification, on existing nuclear power plants before undertaking the large scale importation of generation 3 plants.[46]

This statement clearly did not reflect exactly what NDRC had pursued, as its officials had pushed for importing Generation III reactors with little interest in approving more projects using Generation II reactors. The position taken by the NNSA might be seen as a compromise that those in the nuclear sector would be willing to make – that is, in return for China to agree to import the new type of reactors, they would insist on more projects with their upgraded technology, particularly CNP1000 and CPR1000.

When the international negotiation moved towards the final stages, it was clear that the French would not get the project. Yet, the Chinese side still could not agree whether they would adopt AP1000. Finally, in September 2006, a group of top government and Party officials attended a conference on nuclear technology development. The participants included people from the nuclear industry, the power sector and the machinery industry too. Fourteen out of the 15-member National Leading Group on Nuclear Technology and another 20 experts from various industries were present. After heated debates, 24 supported the adoption of AP1000, 10 supported the two-stage strategy and one abstained. This cleared the way for the final signing of the agreement with the Americans.

On 16 December 2006, a Memorandum of Understanding (MOU) was signed between US Secretary of Energy Samuel W. Bodman and the Chinese minister of the NDRC, Ma Kai. Westinghouse would construct four units of AP1000 in China, two units at Sanmen in Zhejiang and two at Haiyang in Shandong province, a site that was introduced just days before the negotiation closed. The Americans agreed on a full transfer of technology and also accepted China as a partner in the development of the Generation IV nuclear reactor. At the signing ceremony the US Secretary of Energy was quoted as saying, 'the Chinese were very demanding; Chairman Ma and the

NDRC were very demanding' in securing the full technology transfer as a condition for granting the contract. It was not clear if the statement reflected the difficult negotiations or if it was made to calm domestic opposition on both sides of the negotiation – those in the US who were concerned about proliferation issues and those in China who opposed the deal, especially the position taken by the NDRC.

Westinghouse 'agreed to transfer technology to the Chinese that could lead to the construction of many more nuclear reactors in China over the next 15 to 20 years.'[47] Yet, there were conditions for the technology transfer: Westinghouse would transfer technology to Chinese companies, but with the initial supply of four units. This is how France started its own nuclear industry – Framatome was originally licensed to build the Westinghouse model of reactors and it acquired independent control of its technology only in the 1980s. This was also similar to the long-term licensing and technology transfer deal signed between C-E (now part of Westinghouse) and Korean Doosan Heavy Industries (and other Korean companies).

In China, the opposition argued that it was a bad deal because it gave away so many resources and opportunities to foreign companies and placed its own nuclear industry in a position where it would never be able to be independent. Even the Congressional Research Services admitted that 'technological benefits that China might gain from purchasing the Westinghouse reactor ... are likely to be modest,'[48] because AP1000 uses a conventional two-loop PWR, similar to other operating reactor plants in China; the digital instrument and control system is similar to the one used by Siemens-Areva at the Tianwan project; and even with a larger size, the construction would be similar to all other projects in China in terms of welding, pipe manufacture and pressure vessel manufacture.[49]

Westinghouse was a clear winner in signing the deal. It would be able to gain valuable experience in building AP1000 that had never been built before. Eventually this would mean a significant reduction in costs since the cost of the first-of-a-kind is much higher than the projects that follow. The deal also meant that Westinghouse gained a foothold in the Chinese market by squeezing out its competitors, especially the less advanced and less experienced Chinese companies.

Many who had initially criticised the deal finally accepted it because they were told that the Chinese government was under tremendous pressure from the Bush administration to address the trade imbalance between the two countries. China needed large deals like this to make a difference to its trade surplus with the US. With no choice but to accept, CNNC hoped to use the opportunity to develop its own version of the technology based on the imported plant design of AP1000 and prepare for a greater role in the next round of expansion after 2012. Its ultimate goal was to take over, by licence from the vendors, the complete design, manufacture and construction of future nuclear plants. However, this dream was smashed, at least in the eye of the CEO of CNNC, when the State Council announced on 19 April 2007 the creation of the State Nuclear Power Technology Corporation Ltd (SNTPC) to be in charge of technology transfers from foreign countries, including AP1000 from Westinghouse and later EPR from Areva.

In July 2007, SNPTC signed a contract with Westinghouse and the Shaw Group for the 'first-ever deployment of advanced US nuclear power technology in China,' with the contract's value estimated at $8 billion.[50] SNPTC was created as a state-owned corporation with an initial capital of 4 billion yuan in total, coming from the central government budget (60%) and four large state-owned corporations: CNNC, CPI, CGNPC and CTIEC (10% each). SNPTC is authorised by the State Council 'to sign contracts with foreign parties to receive the transferred 3rd generation nuclear power plant technology; to execute the relevant engineering design and project management.' The contract with Westinghouse and the Shaw Group includes the transfer of 'technology in the design and analysis, engineering, licensing, procurement, manufacture, construction, start-up operation, and maintenance of the AP1000 nuclear island.'[51] Some argued that this would require a new institution that would be able to organise domestic enterprises and mobilise resources to 'localise' the capabilities for manufacturing and construction, operation and maintenance, to provide technical support and consulting services as authorised by the government, and later to develop its own brand name nuclear power technology.

In organisational terms, however, SNPTC became a fully fledged competitor and rival, at least to the CEO of CNNC. The rivalry went back to 2004 when the preparatory body of the SNPTC was initially formed as a part of the overall strategy of nuclear expansion in its first stage (before 2010) to: (a) help assess the bids submitted by international vendors for the 3rd generation reactors; (b) monitor the international R&D and (c) develop a team of scientists in the field.[52] At the time, neither CNNC nor CGNPC was happy with the decision to introduce another type of reactor to the industry while they were trying to develop their own competitive technology, CNP1000 and CPR1000. CNNC, in particular, felt its position was threatened because the cornerstone of its technology, CNP models, had been developed out of the SNERDI, the original 728 Institute. Now SNPTC had stripped SNEDRI from CNNC and made it the centrepiece of the organisation.

Even though CNNC and CGNPC each contributed 10% of the initial investment, SNPTC joined the elite state-owned corporations under the supervision of SASAC, on the same level as CNNC and CGNPC. It meant the old two-way competition between CNNC and CGNPC, which was slightly tilted towards CNNC, was replaced with a three-way competition – between Beijing, Shanghai and Guangzhou – for political and administrative supremacy over China's nuclear development. According to an official at the NDRC's Energy Research Institute, SNPTC was created to balance the influence of CNNC and CGNPC. If either CNNC or CGNPC gained control of AP1000, it would mean a monopoly. The industry needed competition and the power and influence of CNNC and CGNPC had to be balanced.[53]

The animosity seemed to be changing after the fall of Kang Rixin, the general manager of CNNC. The new CEO, Sun Qin (孙勤), was trying to build an alliance with SNPTC just two months into the job and signed a strategic cooperation agreement to work on technology transfer, adoption and absorption of GIII reactor technology together with SNPTC. It remains to be seen how this cooperation will work out.

As the main competitor for the Sanmen and Haiyang projects, Areva-ANP lost in its bids to Westinghouse for a variety of reasons: First, Westinghouse was able to make a better offer than Areva on technology transfer partly because it had no significant component fabrication facilities of its own, so it was ready to commit to a much deeper and quicker technology transfer for AP1000 design. It was also prepared to use hundreds of Chinese engineers to do a detailed design for the first Chinese AP1000 units. Areva, instead, had a vertically integrated supplier model with its own manufacturing capability and would do everything to protect its position from competition. Second, after losing the battle with NDRC on developing and expanding its own CNP1000 or CPR1000 reactors, many scientists in the Chinese nuclear industry preferred American technology to French technology because, as Westinghouse had stated, the US had had a better record than France in transferring technology to others. Finally, diplomatic concerns played an important part in the decision and the Chinese government was under huge pressure from the US Congress and the Bush administration to reduce trade surpluses and to appreciate the yuan and was willing to compromise by signing large trade deals such as this to calm nerves in Washington.

In contrast, the offer made by Areva was not nearly as good. Given that Areva-ANP is the joint product of French company Framatome and German company Siemens, in theory any assistance would come from both countries. Yet, Germany could not offer credit guarantees for new nuclear projects due to the policy adopted by the German government in 1998. Though strongly supporting selling EPR to China, the French government could provide such export credit guarantees only by Coface, which was a private company. Coface allowed loans to China of only $325 million, 6.5% of what the US offered, after the French officials decided that China was a good candidate for a Coface guarantee because the country was solvent and the customers had a 'solid' financial structure.[54] The bilateral relationship with France was never on par with the Sino-US relations.

Strangely, however, Areva did not lose the war completely: less than a year later, to the surprise of many in China and the international nuclear community, Areva signed contracts with Chinese organisations to supply two EPRs together with all the fuel and services required to operate them (including uranium supply). The scope of the agreement included establishing an engineering joint venture that would acquire the EPR technology for the Chinese market (ensuring Areva's participation in follow-up projects,) as well as cooperation in the back-end of the fuel cycle, which might lead to the construction of a reprocessing-recycling plant in China. According to IAEA, a contract of this size and scope was 'unprecedented in the nuclear history' of the world.

Like Westinghouse, Areva had strong support from the French government. China had similar problems with Europe as it had with the US – trade surpluses and foreign exchange controls and consequently was subject to the similar pressures from the French government. French president Nicolas Sarkozy publicly made 'the sale of nuclear power to be central to his diplomacy: it was a badge of France's technical prowess and a reaffirmation of its status as a global industrial power.'[55] The deal was signed just before Sarkozy's first visit to China in November 2007. In his formal statement, he thanked President Hu Jintao for his personal involvement in the deal. Yet it was not clear how much Hu Jintao was told about the potential impacts of importing EPR on China's nuclear technology development.

The deal with Areva on 2 EPRs was not only controversial but also bad. It was a bad deal politically because Areva secured the contract without having to compete in an international tender competition for the project, as NDRC had promised in the early 2000s. Some defended the deal because of its fuel component and its financing arrangement – EDF would contribute a one-third share in the operating company, Taishan Nuclear Power Company. But to many, this was a 'dirty' deal. SNPTC eventually signed the agreement with the approval of the State Council, yet it was negotiated by the preparatory body of SNPTC formed by a tripartite team from CNNC, CGNPC and CTIEC. As soon as the deal with Areva was signed, a series of investigations was triggered, but only after the damage to the Chinese nuclear industry had already been done. These investigations eventually brought down the senior officials of all three institutions on corruption charges, taking bribes from foreign companies and using inside information. Kang Rixin (康日新), the president of CNNC, and Jiang Xinsheng (蒋新生), the president of (CTIEC) lost their positions, were expelled from the Party and are waiting for charges to be laid. The deputy president of CGNPC, Shen Rugang (沈如刚), was sentenced to 10 years imprisonment. Many more officials, especially at CGNPC, were either charged or removed from the Party and from their jobs.This was also a bad deal for nuclear development in China. First, instead of concentrating on adopting, absorbing and adapting AP1000 as Generation III technology, the nuclear industry would have to learn EPR and this would only delay standardisation of advanced nuclear technology in China and would create severe difficulties in regulation. Perhaps, this should not be a surprise to the nuclear community because in 2006, the document on nuclear development issued by NDRC for the 11th FYP (2006–10) had already dropped the phrase of 'combining self-reliance with foreign partnership.' In its place, the plan specified that China would introduce Generation 3 reactors from other countries.[56]

Second, there were serious concerns about EPR technology in its first installation at Olkiluoto in Finland. The project ran at least three years behind schedule and €1.5 billion over its initial budget.[57] There was no guarantee that similar delays would not happen in China. Indeed, Framatome ANP, a subsidiary of Areva, formally requested a pre-application review of the EPR reactor design on 8 February 2005, less than three weeks before it put its bid in China. When Areva won the contract from China, the EPR pre-application process had not been completed.[58]

Third, the cost for getting EPR was extremely high: in 2004 the initial bidding was approved by the central government with a price tag of US$1800–1900 per installed kilowatt; the estimated cost for the AP1000 was about US$2300/kw. By the end of 2006, the price tag for EPR had gone up to at least €1800 per installed kilowatt, more than 30% higher than the initial cap,[59] 70–80% higher than US$1300/kw for the domestic technology and design adopted in Qinshan Phase II (CNP600). Finally, this was the first nuclear power project in which a foreign company (a state-owned too) was allowed to be the partner of a new project when it was agreed that EDF could take a 30% share of the company.

One consequence of these policies is that currently all major types of technologies have been used in China's nuclear energy industry and none has been completely localised and standardised (see Table 6.1). This makes it extremely difficult not only to cut down the costs and improve the speed of construction but also to ensure safety standards and to regulate the industry.

Table 6.1 Nuclear power stations under construction in China as of December 2009

	Location	Capacity (MW)	Type of reactor	Starting of construction
Lingao 3, 4	Shenzhen, Guangdong	1080 × 2	CPR1000	December 2005
Qinshan II 3, 4	Haiyan, Zhejiang	650 × 2	CNP600	April 2006
Hongyanhe	Dalian, Liaoning	1080 × 4	CPR1000	August 2007
Ningde	Ningde, Fujian	1080 × 4	CPR1000	February 2008
Fuqing	Fuqing, Fujian	1080 × 2	CNP1000	November 2008
Fanjiashan	Haiyen, Zhejiang	1080 × 2	CNP1000	December 2008
Yangjiang	Yangjiang, Guangdong	1080 × 6	CPR1000	December 2008
Sanmen	Sanmen, Zhejiang	1250 × 2	AP1000	April 2009
Taishan	Taishan, Guangdong	1700 × 2	EPR	December 2009

Note: IAEA uses different labels: those using CPR1000 include: Fangjiashan, Fuqing, Hongyenhe, and Yangjiang; Lingao and Ningde use M310, and Qingshan II 3, 4 use CNP600. CNNC and CGNPC make distinction between the reactors they use: CNNC-CNP and CGNPC-CPR series.

Source: IAEA, 'Power Reactor Information System', at http://www.iaea.org/programmemes/a2/.

The technological development of the Chinese nuclear energy programme is also bad politics because, when there is no single set of consistent policies or a national strategy on nuclear technology development, self-interested enterprises can have a detrimental impact on the whole industry in the long term. They may give the contracts to foreign companies rather than supporting domestic industries in return for personal benefits. For example, a subsidiary of CNNC in charge of the Tianwan project decided to offer the contract of making valves to a California-based company, Control Component Inc., a subsidiary of IMI plc, even though there were Chinese companies that could do the job. In 2009, six former executives of the American company were charged by the US Department of Justice for bribing officials of Chinese state-owned corporations, one of which was Jiangsu Nuclear Power Corp, on the Tianwan project between 1998 and 2007.

When a turnkey project is ordered, major international vendors tend to bring their own partners in designing, manufacturing and building the nuclear power plant. Illustrating the point, Framatome provided nuclear islands, Alstom from the UK provided conventional islands, and German, Swedish and companies from other countries might be used to supply parts or to undertake construction and erection work. This was one of the main complaints from Chinese local industries. The nuclear industry could facilitate overall industrial development in the country if a clear policy is in place. Otherwise, in an environment where foreign companies saw China as a salvation to their nuclear industry and competed fiercely, the Chinese companies had no incentives to help local industries. They often gave foreign companies the contracts to design, manufacture and even install components for a nuclear power plant in the name of securing safety.

Human capital

One key constraining factor on effective technology transfers to developing countries is their 'weak domestic technological capabilities.'[60] Developing human capital takes time and resources. Training sufficient qualified people is more a challenge to China than many developing countries because China would like to transfer foreign nuclear technology to indigenous design and manufacturing and eventually develop its own designs and brands. To achieve these objectives, China will need to train craft labour to build nuclear power plants to international standards, educate engineers for nuclear plant design and train operators for NPPs. China also needs to staff its regulatory agencies with qualified nuclear engineers for oversight and review of new project proposals. This will require significant improvement and expansion of human capital in R&D as well as additional inspections and regulatory capacity.

There is no consensus on the Chinese capacity to adopt, absorb and reverse- engineer imported technologies. Some have emphasised this capacity and therefore warned the West not to sell advanced nuclear technologies to China. After China signed a contract to purchase AP1000 from Westinghouse with technology transfer as part of the deal, for example, some in the US raised concerns about technology transfer: China has already displayed its ability to reverse engineer other militarily useful technologies and with the contract, China could reverse engineer major AP1000 technological advances such as improved, quieter reactor coolant pumps and a digital instrument and control system. This would undermine the competitiveness of American companies.[61]

Others take the opposite view, arguing that even if China imports turnkey projects with an agreement from the vendor on technology transfer, the Chinese do not have the R&D capacity to reverse engineer the technology for military purposes or develop their own brands. The reason for this is partly because the technology is so different – commercial NPPs are not themselves risks for nuclear proliferation. It is also because of the limited human capacity to do reverse engineering, and because of 'poor intellectual property (IP) protection for both foreign and domestic IP development and a poorly developed venture capital industry.'[62] Finally, changed government policies on R&D and especially its R&D funding for tertiary education have undermined China's capacity to standardise and localise advanced imported technology.

In China, CNNC and CGNPC are supported by their own research institutions, some of which are teaching universities while others are design institutes. The research institutes provide technical input. CNNC, for example, has seven research and design institutes under its wing. Much of the researching funding has to come from the operation of these companies. 'This is quite unlike that of the United States, where the prime vendors, such as Westinghouse or General Electric, oversee the design and construction of nuclear power stations and develop the licence application through to final approval by the regulator.'[63] The nuclear technology R&D, however, is by and large funded by the federal government.

In China, there are at least four large bases of nuclear technology development: in the northeast around Beijing, east around Shanghai, southwest around Chengdu and northwest in Gansu. All key research institutes are under the auspices of CNNC, except SNERDI, on which the SNPTC was formed. The most important and the oldest research institute is the China Institute of Atomic Energy (CIAE). CIAE was created in 1950 to prepare for launching the nuclear programme. It was part of the Chinese Academy of Sciences and later placed under the SSTC and COSTIND. It carries out research on all aspects of nuclear science and engineering, for weapons as well as civilian usages, including medicine and agriculture. It hosts several research institutes, such as the Nuclear Power Institute of China, located in Sichuan.

Despite this sprawling network of research institutes, 'China has relied heavily on technology imported from abroad, and the development of its scientific and technological capability has until recently lagged behind its economic growth.'[64] One explanation of this decline was the reform itself. With the reform, (a) the old 'highly bureaucratic and hierarchical R&D structure' was replaced with a much more decentralised and diverse business-supported R&D;[65] (b) government funding for research was increasingly replaced by corporate support (see Table 6.2) and (c) when many public research institutes were converted into business entities, applied research replaced large-scale basic research because corporations wanted to see the immediate results in economic terms.[66]

Moreover, enrolment in nuclear sciences and engineering at universities declined steadily and significantly in the 1990s. Science and engineering were traditionally the favourite fields for university students, while social sciences were shunned because of the political risks – it was much easier to get into trouble with the Party and the government if one was a social scientist, a lawyer or a journalist. Economic reforms offered more opportunities for students in finance, accounting, management, law and other social sciences, and students could easily get jobs in these fields in cities. In addition, most university students enrolled in the 1990s were the only child of the family, and fewer people wanted to work in the remote regions and under the harsh conditions that were often considered normal for people working in the nuclear sector.

Since the mid-1990s, a shortage of skilled workers was noticed in almost all nuclear power station constructions. The Qinshan III construction, for example, had been dogged by a chronic shortage of skilled Chinese workers. 'There is a limit to the amount of skilled labour qualified to do nuclear construction here,' said an executive of AECL's subcontractor. 'There are visible problems in this area at the Qinshan site and (AECL) has to stay on top of this if their 2003 deadlines are to be kept.'[67] To many at AECL, the Lingao projects in Guangdong had attracted many better-qualified workers because it offered higher pay. A shortage of skilled labour was not only daunting for AECL, but could also set back the entire timetable for nuclear expansion in China. As one of the former senior officials at CGNPC put it, 'the most daunting challenge is the qualified labour force.'[68]

Table 6.2   The role of various R&D actors in China, 1987–2004 (% of total R&D expenditure)

	1987	1990	1996	1997	1999	2000	2001	2002	2003	2004
Research Ins.	54.4	50.1	42.8	40.6	38.5	28.8	27.7	27.3	25.9	22.0
Universities	15.9	12.1	11.8	11.3	9.3	8.6	9.8	10.1	10.5	10.2
Enterprises	29.7	27.4	43.3	46.1	49.6	60.0	60.4	61.2	62.4	66.8
Others			2.1	2.0	2.6	2.7	2.1	1.4	1.2	1.0

Source: OECD, 'OECD Reviews of Innovation Policy, China', Paris: OECD, 2008, p.138.

The shortage of qualified university graduates forced CNNC and CGNPC to increase their investment in in-house or joint training programmes with universities to meet the demand. In 1992, CNNC asked Tsinghua University to train those working in nuclear industry for two years to get a second degree in nuclear or electrical engineering. In the following three years, 77 students graduated from the programme. In 1996, CNNC and Tsinghua University signed another contract to train 60 undergraduates in nuclear engineering and nuclear science each year for 10 years. CNNC and Zhejiang University signed a similar agreement to train 40 students each year. Some of these skilled employees were then sent overseas for their training. Between 1978 and 1995, CNNC sent 308 people overseas for postgraduate and other special training and 243 of them eventually returned.

The co-op programmes run by CNNC and CGNPC have been expanding. CGNPC has signed training programmes with nine universities, selecting the third-year university students and providing them with another two years of special training. A couple of conferences were held to deal with the shortage of qualified workers. In 2004, CNNC said it would train 2000 new technicians in three years, and bring in and train about 300 PhDs, 1000 students with master degrees and 10 000 university graduates.

Even with in-house training and enlarged enrolment at universities, the nuclear industry is suffering from the 'nuclear talent vacuum' and the inadequate experience of those working in the field. According to a senior official at the Daya Bay Nuclear Power Operations & Management Co., a subsidiary of CGNPC, the average age of its personnel in 2005 was between 31 and 35 and then it dropped to below 31 years in the following two years. By comparison, the average age of the workforces in the Japanese and South Korean nuclear industry is about 8 and 12 years older, and the average age of the nuclear plant operators is even older. In the US, the average age is over 45 years.[69]

China needs young nuclear scientists and engineers and universities need to attract and cultivate young people in the field for the country's nuclear expansion plan. 'China now needs a batch of young ambitious people to devote themselves to nuclear science, to explore the world of physics,' said Zhu Zhiyuan, director of the Chinese Academy of Sciences, Shanghai branch.[70] One indication of this shortage is that during the recent financial crisis in 2008–09, when the average university placement was a little over 50%, those who had majored in nuclear science and engineering and their related fields had no problems in landing good jobs.

It is too early to tell if Chinese universities can train sufficient staff to satisfy the needs of the nuclear industry, people such as qualified scientists, engineers, operators, managers and regulators. However, the experience in the US shows that it can be done. In the 1960s and 1970s, when the nuclear industry was developing in the US, there were 32 nuclear engineering departments across the country with 1800 students enrolled. This number

doubled in the late 1970s at the peak of nuclear development. When sufficient resources are allocated to universities and it is clear that the nuclear
energy industry is expanding, enrolments will most likely increase in China.
Meanwhile, major nuclear corporations are training or contracting out the
training of qualified people in China as well as overseas. This is nonetheless
a serious challenge the Chinese nuclear industry is facing – training sufficient professional and technical people skilled in operating and managing
nuclear power plants in time to meet the demand.

Conclusion

For three decades, there has been a constant debate on how China would
develop its nuclear energy programme. Few have opposed importing core
technology, while many are in favour of carefully selected imports conditional upon technology transfer. For many, it is not an issue whether China
should introduce foreign technology or cooperate with multinational corporations. Foreign suppliers have already played a significant role in its nuclear
energy development: the involvement of French technology in making fuel
for Qinshan and Daya Bay power plants; the Research Institute for Nuclear
Service Operation, a joint venture with Westinghouse; the training in Spain
of operators for Qinshan, and the German main cooling pumps and injec-
tion pumps at Qinshan, just to list a few. Yet there has been deep suspicion
about the willingness of these multinational corporations to share technology because, on several occasions, they agreed in the contracts to do
so but by the time the contract went into operation the technology had
already been abandoned by the companies.[71] 'Market exchange will never
bring you the most advanced core technology', claimed Chinese scientists
and engineers.

The main reason China has not been able to achieve what Japan and
South Korea have done in building up a nuclear industry is more about politics than technology. In China, there is not a set of coherent national strategies that have supported and encouraged basic research and development in
nuclear science and engineering, as was the case in Japan and South Korea.
Nor have there been consistent policies in guiding the technology development, adoption and adaptation. As many scholars have pointed out, both
Japan and South Korea managed to grow quickly into industrialised economies to a large extent because of their governments' industrial policies that
provided targeted and subsidised credit and public investment in both basic
and applied research and development, the pervasive state administrative
guidance and entrepreneurship, and an overarching 'reciprocal principle' of
never giving anything to businesses for free without stipulating a monitor-
able performance standard in exchange. In particular, it was their focus on
production engineering, not cheap labour that allowed both countries to

start the nuclear industry with imported key technologies and then build their niche.[72]

In China, it is the absence of this set of consistent industrial policies that has led to what many in the nuclear sector have called the 'united nations' approach – importing technologies from many countries without being able to absorb, localise or standardise any of them. David Nobel argued some time ago that science and technology policy was far from a value-free enterprise. Science typically helps legitimise and reproduce the power of the dominant political and economic interests in society.[73] The absence of a set of consistent policies for nuclear technology development in China reflects the politics of China today – bureaucratic infighting and competition among various levels of government. Rivalries within the nuclear industry and governments can explain the erratic policies and their adverse impacts on nuclear technology development.

Notes

1. Lewis, 2007, p.232.
2. Evans, 1995, p.15.
3. US Congress, Office of Technology Assessment, *Energy Technology Transfer to China: A Technical Memorandum*, OTA-TM-ISC-30. Washington, DC: US Government Printing Office, September 1985, p.36.
4. Fewer and Altvater, 1977; Shin, et al., 2007.
5. Evans, 1995, p.72.
6. Simon, 1987, p.249.
7. 'China's New Priorities for Technology Development', *China Business Review*, 5:3, May–June 1978, pp.3–8.
8. Quoted in Jones, 1981, p.35.
9. Simon, 1987, p.249.
10. US Congress, Office of Technology Assessment, *Energy Technology Transfer to China: A Technical Memorandum*, OTA-TM-ISC-30. Washington, DC: US Government Printing Office, September 1985.
11. Price, 1990, p.51.
12. Ibid, p.52.
13. Theresa Tan, 'Daya Bay: China's New Powerhouse', *Strait Times*, 6 March 1994; Mark Hibbs, 'China Operates its Largest Nuclear Units: success of Daya Bay PWRs will Rest on China's Talent Scouting', *Nucleonics*, 35:7, 28 April 1994, p.12.
14. Park, 1992.
15. Quoted in Jones, 1981, p.35.
16. Ibid.
17. Han Guojian, 'China: A Country of Nuclear Power', *Beijing Review*, 23–29 December 1991, p.12.
18. Lieberthal, 2004, p.77.
19. Margaret L. Ryan, 'Work Proceeding, but Slowly, on China's Indigenous Nuclear Plant', *Nucleonics Week*, 28:42, 15 October 1987, p.12.
20. Quoted from Ann MacLachlan, 'Engineering Advances for Next PWRs at Qinshan', *Nucleonics*, 36:7, 27 April 1995, p.8.

21. 郝冬琴, 加快我国核电发展的基础条件和几点设想, 核经济研究, (Hao Dongqin,, 'Conditions and Recommendations on China's Nuclear Energy Expansion', *Nuclear Economic Research*), 3, 1996, pp.21–25.
22. 孙敏莉'中俄战略合作的重大成果', 中国核工业, (Sun Minli, 'Sino-Russian Cooperation', *China Nuclear Industry*), December 2007, pp.16–17.
23. 张欣 '田湾核电站: 中俄合作的成功结晶' 中国核工业, (Zhang Xin, 'Tianwan Nuclear Power Station: The Success of Sino-Russian Cooperation', *China Nuclear Industry*), December 2007, pp.26–28.
24. 路风, '被放逐的中国制造: 破解中国核电谜局', 商务周刊, (Lu Feng, 'Exiled Chinese Manufacturing Industry: Demystifying China's Nuclear Industry' *Business Weekly*), 2, 2009, pp.30–54.
25. William M. Spodak, 'Power Struggle', *The China Business Review*, 25:2, March/April 1998, p.29.
26. 核电发展战略研究编委会, 1999, pp.9–11.
27. Mark Hibbs, 'Beijing Order Nuclear Sector to Revamp PWR Development Plan', *Nucleonics*, 41:1, 6 January 2000, p.4.
28. Zhidong Li, 'The Prospects for Nuclear Energy in the East Asian Region: Focusing on China', *International Journal of Global Energy Issues*, 30:1/2/3/4, 2008, p.274.
29. Kan and Holt, 2007, p.3.
30. Kevin Platt, 'China's Nuclear-Power Program Losses Stream', *The Christian Science Monitor*, 21 July 2000.
31. Kan and Holt 2007, p.19.
32. EIA, 'Future of the Chinese Nuclear Industry', at http://www.eia.doe.gov/cneaf/nuclear/page/nuc_reactors/china/outlook.html, accessed on 10 November 2009.
33. 'US-China Agreement Provides Path to Further Expansion of Nuclear Energy in China', news release by the US Department of Energy, December 16, 2006 at http://www.energy.gov/print/4536.htm.
34. Shirley Kan, 'US-China Nuclear Cooperation Agreement'. CRS Report to Congress, 6 September 2007, p.20.
35. Ibid.
36. 郭丽岩, 王岩敏, '国家自主创新战略的核心要义: 重大装备本土化', 改革与战略, (Guo Liyan and Wang Yanmin, 'The Importance of National Innovation and Localisation', *Reform and Strategy*), 172:12, 2007, pp.1–4, 15.
37. '30亿元市场蓄势: 国产核阀需大步前行'机械工程师, ('3 Billion Market', *Engineering*), 7, 2007, pp.9–10.
38. 李永江, '关于对我国核电主设备制造进行宏观引导和布局的建议'中国核工业, (Li Yongjiang, 'Suggestions on Self-Reliance on Nuclear Equipment Manufacture', *China Nuclear Industry*), 79:3, 2009, pp.15–16.
39. Mark Hibbs, Ann Maclachlan and Steven Dolley, 'With Governments Supporting Bids, Competition in China Looks Fierce', *Nucleonics*, 46:8, 24 February 2005, p.1.
40. See discussions in Anonymous, 'Concerns about Hollowing China's Nuclear Energy' and other articles, April, 2006 at http://caini.cnnc.com.cn/zhuanti/06-guochanhua/index.htm.
41. Quoted from '中国核电发展: 走在自主与引进的平衡木上', 中国企业家, ('Chinese Nuclear Development: Walking on a Balance of Self-Reliance and Imports', *Chinese Entrepreneurs*, at http://www.cnnc.com.cn/Portals/0/zhuanti/06-guochanhua/P2.htm
42. Parmy Olson, 'Nishida's Toshiba Wins Westinghouse With $5.4B Bid', *Forbes*, February 6, 2006.

43. Mark Hibbs and Ann MacLachlan, 'Vendor Officials in China Won't Deny Rumours PWR Bidding is Over', *Nucleonics*, 47:13, 30 March 2006, p.1.
44. Mark Hibbs, 'CNNC Battling with State Council over Direction of China's Program', *Nucleonics*, 46:24, 16 June 2005, p.14.
45. http://unpan1.un.org/intradoc/groups/public/documents/APCITY/UNPAN007903.pdf.
46. Ganjie Li, Xiaofeng Hao and Bo Tang, 'Nuclear Safety Regulation and Review of New Nuclear Power Plants in China', in IAEA, *Effective Nuclear Regulatory Systems: Facing Safety and Security Challenges*, Vienna: IAEA, 2006, p.255.
47. 'Statement by US Secretary of Energy Samuel W. Bodman on Reaching Agreement on Civilian Nuclear Energy with China', http://www.doe.gov/news/4537.htm.
48. Kan and Holt, 2007, p.22.
49. Stephen V. Mladineo and Charles F. Ferguson, 'On the Westinghouse AP1000 Sale to China and Its Possible Military Implications', Non-proliferation Policy Education Centre Research Memorandum, 11 March 2007, http://www.npec-web.org/Essays/20070311-MladineoFerguson-WestinghouseSaleToPrc.pdf.
50. Westinghouse, news release, 'Westinghouse, Shaw Group Sign Landmark Contract to Provide Four AP1000 Nuclear Power Plants in China', 24 July 2007; and Chen Aizhu and Jim Bai, 'Westinghouse Seals Mega China Nuclear Deal; *Reuters*, 24 July 2007.
51. Andrew C. Kadak, 'China's Energy Policies and Their Environmental Impacts'. Testimony before the US-China Economic and Security Review Commission, 13 August 2008.
52. Anonymous, 'China Sets Up State Nuclear Power Technology Co.', May 22, 2007, http://english.peopledaily.com.cn/200705/22/eng20070522_376921.html.
53. 余力，康日新冲击波，南风窗，(Yu Li, 'Kang Rixin Attack', *Window of Southern Wind*), 7 September 2009. Mark Hibbs, 'China's Corruption Probes at CNNC May Be Expanding to Subsidiaries', *Nucleonics*, 50:33, 20 August 2009, p.1.
54. Mark Hibbs, Ann Maclachlan and Steven Dolley, 'With Governments Supporting Bids, Competition in China Looks Fierce', *Nucleonics*, 46:8, 24 February 2005, p.1.
55. 'Power Struggle,' *Economist*, 4 December 2008, pp.81–82.
56. http://stxx.costind.gov.cn/n435777/n711341/n711753/n801764/appendix/200741384442.doc.
57. 'Power Struggle,' *Economist*, 4 December 2008, pp.81–82
58. 'Report of the Nuclear Energy Committee', *Energy Law Journal*, 28, 2007, pp.767–83.
59. Ann MacKachlan, 'Areva Hits Jackpot with Contract for Islands, Fuel for Taishan EPRs', *Nucleonics*, 48:48, 29 November 2007, p.1.
60. Gallagher, 2006, p.384.
61. Stephen V. Mladineo and Charles D. Ferguson, 'On the Westinghouse AP1000 Sale to China and its Possible Military Implications', Non-proliferation Education Centre Research Memorandum.
62. Lampton, 2008, p.136.
63. Andrew C. Kadak, 'Nuclear Power: "Made in China"', MIT (2006), http://web.mit.edu/pebble-bed/papers1_files/Made%20in%20China.pdf.
64. OECD, 2008, p.32.
65. Sylvia Schwaag Serger and Magnus Breidne, 'China's Fifteen-Year Plan for Science and Technology: an assessment', *Asia Politics*, 4, 2007, p.138.
66. Denis Fred Simon, 'Science and Technology Reforms', *China Business Review*, 12:3, 1985, pp.31–35.

67. Mark Hibbs, 'Candu in China on Schedule Despite Skilled Labour Drain', *Nucleonics Weekly*, 41:13, 30 March 2000, p.6.
68. 冯志卿, '中广核以核养核', 中国投资, (Feng Zhiqing, 'CGNPC's Rolling Development', *China Investment*), March 2008, pp.66–70.
69. Mark Hibbs, 'Potential Personnel Shortages Loom Over China's Nuclear Expansion', *Nucleonics Weekly*, 49:24, 12 June 2009, p.1; Andrew C. Kadak, 'US-China Energy Technology Cooperation: Civil Nuclear Energy', Testimony before the US-China Economic and Security Review Commission, 13 August 2008.
70. 'China Energy', *China Daily*, No. 1, 1–31 March 2009, p.34.
71. Li Er-kang, 'Commenting on the Path of Self-Reliance in Nuclear Development', July 28, 2006 at http://www.atominfo.com/cn/forum/printer_freindly_posts.asp?FID=1&TID=1594.
72. Amsden, 1989; Johnson, 1982; World Bank, 1993.
73. Noble, 1977.

7
Fuelling the Future: The Nuclear Fuel Cycle

Nuclear fuel supplies have two-dimensions: uranium resources – its exploration and prospecting – and the nuclear fuel production and services that allows uranium to be converted into fuel to feed reactors. Each of these two aspects raises different concerns and challenges. There is anxiety about whether the world's uranium reserves can sustain the rapid expansion of nuclear energy programmes. The nuclear fuel cycle consists of the front-end and back-end. The front-end of the nuclear cycle consists of uranium exploration, mining and milling, conversion, enrichment and fuel fabrication. The back end consists of spent fuel disposal, reprocessing and storage of high uranium waste. Currently, only a few countries in the world have full nuclear cycle capability and only a very few run large commercial nuclear fuel services (Canada, France, Russia, the UK and the US for conversion; France, Germany, Netherlands, Russia, the UK and the US for enrichment; and France and the UK for reprocessing services).

Securing nuclear fuel supplies raises three sets of challenges to the Chinese nuclear industry: First, China hosts only 1% of the world's identified uranium resources. Expansion of its nuclear energy programme requires more investment in domestic uranium exploration and prospecting. Chinese firms are also encouraged by the government to invest in overseas uranium exploration and mining. The world's uranium industry is dominated by a few global conglomerates and the few places where there are still opportunities for Chinese investment are high-risk and highly unstable places. Chinese nuclear companies need to identify and purchase uranium resources abroad and seek production opportunities for those resources. They can only do so by avoiding direct competition and potential confrontation with major powers, and therefore face some serious challenges when investing in high-risk countries both in economic and security terms.

Second, China has built a front-end nuclear fuel cycle capacity, but expansion of a nuclear energy programme would require a significant increase in nuclear fuel production. China is a nuclear state and has placed its nuclear energy programme under the safeguard provision of IAEA. This has reduced risks in turning a civilian to a weapons programme. Its nuclear fuel production and enrichment facilities nonetheless are safety and security concerns.

Third, given its limited uranium reserves, Chinese scientists have been working on the back-end of the nuclear fuel cycle – reprocessing capacity and fast breeder reactors. The policy of the country is to become self-sufficient in nuclear fuel supplies and to make nuclear fuel recyclable and the industry sustainable. This is a technical challenge, as well as an economic and political change. Developing an integrated nuclear fuel cycle technology and capacity is expensive and it is difficult to obtain the technologies necessary to carry out the development. It has the potential to heighten proliferation concerns in the world, and it will create another set of nuclear waste issues – for example, smaller quantity but with higher levels of radioactivity.

Nuclear fuel production in all countries is handled by the state in secret, predominantly because, once countries have mastered uranium enrichment and plutonium separation technologies, they are nuclear weapon *capable* states that have the potential to develop nuclear weapons within a short time. According to the IAEA Director General, Dr Mohamed ElBaradei:

> This is too narrow a margin of security, in my opinion. These countries may have no intention of ever making nuclear weapons, but that can change quickly if their perception of the risks to their national security changes. And security perceptions, as we know, can change very rapidly.[1]

This is the reason the international community has strict rules regulating nuclear fuel services and has also been working on a multilateral approach to nuclear fuel supplies. How to encourage market competition, while maintaining tight regulation on safety and security, is the challenge facing all countries. China supports the multilateral measures but prefers to be on the supply-side.

This chapter examines the development of these three aspects of nuclear fuel supplies – uranium exploration and mining, fuel production, and development of the back-end of the nuclear cycle capacity. In all three aspects, we will find the repeated message in this book – nuclear energy development may be pursued to ensure energy security, the lack of consistent policies in China can explain some of the failures in securing the nuclear fuel supplies.

Nuclear fuel supplies

Even though uranium is a relatively common element of the earth's crust, China has limited identified uranium reserves (reasonably assured resources, RAR, and inferred), and about 1% of the world's total RAR uranium. China had its first uranium discovery in 1954, and between 1954 and 2007, it invested more than 11 billion yuan in uranium exploration. The real progress was made in its early years. As in the coal and oil sector, in the 1980s uranium was exported in order to obtain badly needed foreign exchange for the country. 'The first Chinese long-term [uranium] contract with a US utility was signed in 1988, when China's export potential was believed to be 700–1500 metric tonnes U per year.'[2] Its exports were short-lived, first as the result of political instability and then because the whole domestic industry suffered as the government's budget support and public procurement were cut considerably.

In the 2000s, the desire to expand its nuclear energy generation capacity has challenged the domestic uranium industry. According to Chinese officials, if China expands its current nuclear generation capacity of 9.1GWe to 40GWe by 2020, the accumulated requirement for uranium would reach 10 867 tonnes of uranium (tU) by 2010, 40 300 tU by 2015 and 87 047 tU by 2020. Currently, according to NEA, China has RAR of 32 000 t at US$40/kgU and 49 000 t at US$130/kgU.[3] It produces about 840 tonnes of natural uranium each year now, which by 2020 would not be enough even for one year's consumption. Given the high costs of domestic exploration and production, the Chinese government has adopted a three-way strategy to meet domestic demand – a combination of domestic production, imports and production from Chinese-invested foreign sources. This is the strategy written in the Medium- to Long-Term Nuclear Energy Development Plan (2005–20), a combination of: (a) domestic production; (b) 'going out' to invest in overseas uranium mines and (c) trade in (purchasing on the international market) – 建立国内生产, 海外开发, 国际铀贸易三渠道并举.

Domestic uranium exploration and mining

Domestic uranium exploration and mining suffered greatly from the late 1980s to about 2002, partly because of the international developments and mainly because of domestic politics. Globally, the 1980s and 1990s saw a steady decline in uranium exploration and mining as the uranium price on the world market declined steadily from its peak US$243/kgU in 1977 to US$18/kgU in 2000 (in 2003 dollars).[4] The prolonged period of low uranium prices led to a steady decline in investment in the industry, which in turn 'led to the closure of all but the lowest-cost mining facilities, stimulated market consolidation and curtailed investment in exploration and mine development.'[5] In the same two decades, the uranium requirements exceeded its production and the gap was met largely by the secondary supplies – uranium inventories accumulated in the previous years and converting weapon-grade enriched uranium.

In China, the industry suffered as the central government cut its budget, as its export markets closed down, and as it was unable to reduce its redundant labour force. Uranium exploration was labour-intensive and its workforce accounted for about 25% of all employees at CNNC. In 1995, CNNC initiated a reform of its uranium exploration by closing 10 mines and improving performance. As other sectors in China improved rapidly in their productivity and living standards, the uranium industry moved backwards, accumulating huge losses. At the end of 1998, under CNNC, there were six regional uranium geological bureaus, spread across 26 provinces and regions. The total number of employees was 64 166, but 48% of them were retirees that CNNC had to support. The rest included not only those working in uranium prospecting and exploration but also those who worked in hospitals, schools, research institutes and other facilities. These social functions had been an integral part of the operation because teams were often located in deserts or remote areas where there were few opportunities for the employees and their families and everything had to be provided by the work units. Consequently, uranium prospecting and exploration of the CNNC was its heavy loss-making segment, with losses running to 300 million yuan between 1996 and 1998 alone.

In the second half of the 1990s, there were two opposite developments in China: 'the decrease of uranium exploration activities from the mid-1990s to the end of the decade' and the 'speed up [of] the construction of nuclear power plants in coastal areas.'[6] These were the direct consequences of the policies adopted by the central government in the mid-1990s, especially after 1997 when Zhu Rongji took over as the premier of China. He made it clear that CNNC's ambition to add 10GWe of additional nuclear power plants in the coming decade would be rejected unless it addressed the loss-making problems. To turn around the situation and, more importantly, to protect the nuclear industry as a whole, those in charge of restructuring the institution decided to hive off its non-profitable segments of uranium exploration and mining, and transfer them to the National Bureau of Geology. The industry was nearing collapse.

In April 1999, the State Council approved this proposal and issued a document, the Reorganisation Plan of Geological Exploration (地质勘查管理体制改革方案), which stated that CNNC would keep a small team of personnel on radioactive material exploration, and the majority of the personnel involved in geological exploration and prospecting would first be transferred to National Bureau of Geology and then to provinces. The provinces were not allowed to transfer them down to lower levels of government. CNNC transferred 30 011 out of its 33 149 regular employees (90%) and 28 684 out of its 31 017 retired people (92%) to the National Bureau of Geology which soon cut its employees from 45 000 to only about 5500, with the whole uranium exploration team being transferred to provinces.[7] CNNC would keep 3138 regular employees and 2333 retired people as its core team engaging in uranium prospecting and exploration activities.[8] The State Council ordered the transfer to be completed in a year. In the following five years, however, CNNC continued to pay a large amount of money to settle those transferred and especially those laid off in one form or another.

At the same time, the number of people working in the nuclear fuel sector was cut too, and more than 21 000 people were asked to leave.[9] To keep the number of employees under control, CNNC adopted a policy in 1998: it would deduct 100 000 yuan from its exploration and mining budget for every additional person to be brought on board and provide a 'reward' of 30 000 yuan for every person to be dismissed. This 'strange' incentive structure was set up mainly in response to the pressures from the employees – their remote location meant very limited job opportunities for their children, who used to get on the payroll as a regular practice.[10] This policy caused resentment not only among the employees but also among some veterans in the nuclear industry, who demanded that the central government do more to improve the living conditions for people who had made contributions to Chinese nuclear development. The plea fell on deaf ears of Zhu Rongji.

In 1999, a major reorganisation was adopted. The old CNNC was split into two – a new China National Nuclear Corporation that included uranium exploration, mining and milling, other aspects of nuclear fuel production, and the nuclear energy industry itself, and the China Nuclear Engineering and Construction Corporation, which would be in charge of NPP construction. Under CNNC, the China Nuclear Energy Industry Corporation (CNEIC) was created as its subsidiary, to be in charge of uranium mining and enrichment services. With the restructuring in 1999, provinces took over the assets, debts and personnel in geological exploration, including those in uranium prospecting and exploration.

In 2003, the State Council allowed the geological exploration teams that had gone through the corporatisation process to lease or transfer the land-use rights and the right of land-use management. Provinces where these teams were located benefitted directly from these transfers, which let them boost uranium exploration at the time of a rising shortage of electricity supplies. In the provinces where uranium was discovered, uranium reserves were used as a bargaining chip with the central government for the approval of NPPs. Jiangxi province, for example, has 1.3% of the national hydro reserves and 0.137% of the national coal reserves, but one-third of the country's known discoverable uranium reserves, two-and-a-half times the reserves of Guangdong and four times the reserves of Zhejiang. The prevailing view was that since both Guangdong and Zhejiang have had their NPPs, Jiangxi should have its share too.[11] The same argument has been made by other provinces, such as Hunan and Guangxi, to have their potential sites for nuclear power plants approved in these inland provinces.

Opportunities for uranium exploration and mining came in the new millennium. After the IPCC report was issued in 2000, and with rising energy prices from 2003 onward, nuclear power became an attractive option for many countries, including China. The world's uranium price rose steadily, from US$18/kgU in 2000 to US$52/kgU in early 2005 and then from US$351/kgU in mid-June 2007, an almost 20-fold increase in only seven

Table 7.1 Industry and government uranium exploration and development expenditures – domestic

	2000	2001	2002	2003	2004	2005	2006
US$ million	115.2	89	95.1	123.8	218.8	364.1	773.8
% of previous year		77.2	106.9	130.2	176.7	166.4	212.5

Source: NEA/IAEA, *Uranium 2007: Resources, Production, and Demand*. Paris: OECD, 2008, p.32.

Table 7.2   The distribution proportion of known reserves

	Europe	Australia	North America	Asia[a]	Asia[b]	Africa	Russia	South America	Northwest China
RAR allocation (tU/km^2)	0.1407	0.1252	0.0676	0.0498	0.0124	0.0372	0.0226	0.0156	0.0055

Notes: a – including Kazakhstan, Uzbekistan, Tajikistan and Mongolia
b – not including Kazakhstan, Uzbekistan, Tajikistan and Mongolia

Source: 叶柏庄，'我国铀地勘业可持续发展的战略思考', (Ye Baizhuang, 'Sustainable Development of China's Uranium Resources', *Advances in Earth Science*), 21:11, 2006, p.1137.

years[12] (see Table 7.1). Higher prices brought in more investment in uranium exploration. 'A very significant increase in exploration and development activities occurred in 2005 and 2006, driven by increases in the uranium spot price.'[13] In 2005, 19 countries reported domestic exploration and development expenditure totalling about US$364 million, an increase of about 66% compared with 2004.

In China, it became clear that the neglect of the uranium industry had to be reversed if the country insisted on 'feeding' its expanding nuclear generation capacity with its own uranium. To many insiders, uranium reserves were as widespread in China as they were in many other countries (see Table 7.2). The difficulty of uranium exploration and mining is three-fold: (a) it normally takes 10–20 years from exploration to mining; (b) it takes at least US$8 million to find a regular uranium mine and (c) it takes new and advanced technology to do so in a country as vast and as diverse as China. According to those in the industry, the low levels of known reserves in China did not necessarily mean that China did not have the reserves; it meant that the country had not poured enough investment into the industry.[14]

The call for an expansion of nuclear energy programmes in China has led to an increase in investment in uranium exploration, prospecting and mining. In the 10th FYP (2001–05), CNNC invested 3 billion yuan in uranium mining. From 2003 onward, the government expenditure on uranium exploration rose from US$7.2 million in 2003 to US$9.5 million in 2004, US$13.5

million in 2005 and US$25.5 million in 2006, tripling that of 2003. It was also expected to increase by more than 30% to US$33.6 million in 2007.[15] With increased investment, activities expanded. Some concepts discussed by scientists in the late 1990s were picked up – that is, China would have to shift its exploration from hard-rock hosted targets mostly located in south-east China, to sandstone-hosted uranium deposits in northwest China.[16] As 'investment in uranium exploration steadily increased from the year 2000, drilling distance experienced a rebound from 40 000m to 70 000m in 2000, gradually increasing to 130 000m in 2003 and 140 000m in 2004.'[17] In 2006 and 2007, the total drilling distance increased to 600 000m. 'As a result, significant discoveries of uranium resources in northern China added more than 15 000tU of Identified Resources.'[18]

In 2005, the State Council adopted the Framework for Developing the Country's Uranium Resources (我国天然铀资源发展规划纲要) which stated that uranium exploration should be open to other organisations in addition to CNNC. Lowering the entry barrier was necessary because the investment in uranium exploration, prospecting and mining tended to be high and the CNNC did not have sufficient financial resources to invest in uranium exploration in large areas. Yet, this opening-up had to be done in an ordered way to avoid huge waste and losses, as it happened in China's coal industry.[19] On 12 February 2006, the State Council issued a No.4 document on strengthening geological exploration. This was the first time ever that the government had issued its official policy on uranium resources – 'strengthening uranium exploration, speeding up its new exploration'.

In 2008, the Ministry of Land and Resources issued a document on uranium exploration and mining. It emphasised the cooperation between the CNNC's uranium bureau and geological teams in provinces; it called for an increase in the budget allocation to the industry from the central government; and it also made it clear that other investors should be encouraged to invest in uranium exploration and mining based on the principle of 'whoever invests benefits' – the principle adopted in the power sector in the 1980s that set in motion a rapid expansion of electricity generation capacity in China.[20]

In 2008, CNNC signed two agreements with Qinghai province and the Xinjiang Autonomous region to cooperate on resource exploration. It was expected to sign similar agreements with Gansu and Inner Mongolia. Despite these efforts, many in the field still argue that domestic uranium reserves fall far short of meeting the rising demand and, indeed, the gap is quite large. According to one estimate, if China maintains the current level of uranium production of 1200tU per year, the gap by 2020 would be about 80% (see Table 7.3).

According to another calculation, even if China significantly increases its uranium production, the gap in meeting the demands by 2020 remains large (see Table 7.4).

Table 7.3   Analyses and prospective of China reactor and related uranium requirements

	2002	2003	2005	2010 Low	2010 High	2015 Low	2015 High	2020 Low	2020 High
Nuclear generation capacity (MW)	4400	6100	8700	12,700	14,700	13,900	15,600	13,900	15,600
Uranium Requirement (tU)	790	1100	1570	2290	2650	3240	4140	3960	5760

Source: 叶柏庄, '我国铀地勘业可持续发展的战略思考', (Ye Baizhuang, 'Sustainable Development of China's Uranium Resources'), *Advances in Earth Science*, 21:11, 2006, p.1136.

Table 7.4   Natural uranium requirements and production (2010, 2015 and 2020)

	Requirement/ tonnes	Production/ ton	Production gap/ton	Demand/supply ratio %
2010	2971	2180	−790	73.4
2015	4929	3030	−1899	61.5
2020	8769	3870	−4899	44.1

Source: 戴民主, '江西省核电发展与铀资源保障', 江西能源, (Dai Minzhu, 'Nuclear Development and Uranium Supplies in Jiangxi', *Jiangxi Energy*), 4, 2008, p.3.

Chinese officials at international conferences have stated that the domestic production of natural uranium would be able to meet the demand with its expansion of nuclear capacity to 40GWe by 2020. Nonetheless, many in the industry, including IAEA and OECD Nuclear Energy Agency (NEA), have raised serious doubts and called for more investment in uranium exploration, mining and milling. Concern about inadequate uranium supplies is one of the reasons some people have called for a cautious expansion of the nuclear energy programme. For others, this is exactly the reason for adopting a 'going out' strategy – purchasing equity uranium stakes abroad that would allow for ownership of the resources 'in the ground' – to meet rising domestic demand.

'Going out'

In the early 2000s, before the uranium industry had recovered from the reorganisation, the demand for uranium was rising in China. Predictions emerged that domestic production of uranium would not be able to meet the need if nuclear generation capacity expanded. The world's identified uranium resources (5.55 million tonnes in 2007) are concentrated in a few countries:

Australia (22%), Kazakhstan (15%), Russia (10%), Canada (8%), South Africa (8%), Brazil (6%), the US (6%), Niger (5%) and Namibia (5%); together they account for 85% of the world's total reserves. Some of these countries such as Australia, Kazakhstan, Niger and Namibia do not have nuclear power generation capacities at all. Uranium is a true global market. IAEA estimates that 'identified uranium resources are sufficient to fuel an expansion of global nuclear generating capacity, without reprocessing, at least until 2050.'[21]

Neither the availability of sufficient uranium reserves nor existence of a global uranium market has calmed the anxieties of many in China. For them, with only 1% of the world's total uranium reserves, China must invest in overseas uranium exploration and mining to meet its future needs. Their arguments are both economic and geopolitical. The world uranium industry is dominated by a few conglomerates with huge resources at their disposal. By 2007, eight major companies controlled about 86% of the global production with 52% controlled by the three largest producers alone: Cemeco, AREVA and Rio Tinto. Merger negotiations between Rio Tinto and BHP Billiton in 2008–09 triggered a lot concerns among Chinese industries and policy makers not only because of their control over iron ore or coal but also uranium, with Rio Tinto controlling 61.5% of the world's total production and BHP Billiton (5.9%).

There is also a long-term relationship between sellers and buyers. About 90% of the world's uranium is sold and bought on long-term contracts directly between producers and buyers, which are often nuclear operation companies, in the name of guaranteed un-interrupted operation. This meant that much of the current uranium production had already been committed, and thereby it would be difficult for China, as a new comer, to find sellers. Despite these global market situations, it would make economic sense for Chinese companies to look into the overseas investment opportunities because the marginal cost of exploring Chinese resources was on average considerably higher than for foreign production.

In 2001, in the 10th FYP, the central government included 'going out' as one of the 'four key thrusts to enable China to "adjust itself to the trend of economic globalisation".'[22] In 2002, Zen Peiyuan, then vice-premier of the State Council in charge of energy policy, said China should open up two uranium markets: domestic and international. Domestically, the government should put more resources into uranium exploration, mining and milling, and internationally, its corporations should go out and invest in overseas uranium exploration and mining. This was an extension of the policy adopted by the central government in the early 1990s where China would supplement its early policy of 'opening up' by attracting foreign investment to China with encouragement of 'its enterprises to expand their investments abroad and their transnational operations.'[23]

In 1997, Jiang Zemin reemphasised the importance of encouraging SOEs to form highly competitive large enterprise groups with transnational activities.[24] The policy was adopted partially for domestic reasons – turning many large loss-making SOEs into commercial entities. It was also adopted because 'increasingly, the Chinese want to capture a greater portion of the "value chain" in the production of goods, no longer concentrating on providing low-cost labour (what the Chinese call *jiagong* or "adding labour") to assemble products.'[25]

Chinese energy companies, particularly its oil companies, led the shift: China National Petroleum Corporation (CNPC) began investing abroad in the early 1990s. It 'purchased reserves in Canada in 1992, signed a production-sharing contract in Thailand, successfully bid to improve oil recovery at a Peruvian field in 1993, and signed an agreement to explore oil in central Papua New Guinea in 1994.'[26] By 1997, it was widely accepted in China that overseas investment was necessary to secure the country's energy supplies, which had become increasingly dependent on imports.

> Unless China invests the capital to control some oil resources, any even insignificant international economic, political, or military conflict could affect the supply and demand on the spot market, causing severe interference to our oil imports, to seriously undermine China's economic stability and sustained development.[27]

Chinese large state-owned corporations were encouraged to invest in Africa, Latin America and Central Asia, not only in resources but also in other industries. China might or might not have developed a 'grand strategy' on its 'going out' campaign; its large firms had an immediate interest in gaining access to the overseas resources for their expansion. This was also the case with the nuclear industry. CNNC carried 'an air of improvisation, if not desperation, and [was] an odd blend of shop-worn party sloganeering ('Go Out!'), and 21st Century State-directed capitalism.'[28] This exercise of investing in uranium exploration, prospecting and mining, however, took place in an environment where a few conglomerates dominated the field.

To search for and get access to global uranium resources, the State Council streamlined and consolidated prospecting operations. In 2004, China Nuclear International Uranium Corporation (Sino-Uranium) was created as a subsidiary of CNNC to invest in overseas uranium exploration and mining. Sino-Uranium was given 50 million yuan as initial capital for operation and it had only six people working for the corporation. With the back-up of CNNC, Sino-Uranium quickly became a key player in overseas investment. It collaborated with other large state-owned corporations, supported by the state-owned policy banks, targeting regions such as Africa and Central Asia. The Chinese 'going out' policy coincided with an increase in investment by other major countries (see Table 7.5). Competition in Africa and Central Asia among large uranium mining companies, nuclear power states such as Japan and South Korea, and the Chinese firms was tense.

Table 7.5 Non-domestic uranium exploration and development expenditures (US$1000 in year of expenditure)

Country	2000	2001	2002	2003	2004	2005	2006	2007 exp.
Australia	NA	NA	NA	NA	1571	8855	4580	4724
Canada	3667	2597	2549	2547	9559	53,968p	12,4546p	139,655p
France	7330	7690	14,370	16701	59,701	127,500	8500	115,000
Total	10,997	10,287	16,919	19,248	70,834	190,323	214,129	259,395

Notes: Domestic exploration and development expenditures represent the total expenditure from domestic and foreign sources within each country. Expenditures abroad are thus a subset of domestic expenditures.
p – provisional data

Source: NEA/IAEA, *Uranium 2007: Resources, Production, and Demand*. Paris: OECD, 2008, p.30.

In Africa, the Chinese policy was described by a senior official at Sino-Uranium: 'Resource-abundant countries give us resources, and we help them build infrastructure.'[29] Initially, the battle in Africa was between CNNC and the Russians over the uranium reserves in Namibia, Niger and South Africa and then between CNNC and the French Areva. When the global uranium price jumped from US$18/kg in 2000 to US$52/kg in 2005, the vice-president of CNNC said that 'CNNC was alarmed by the very quick rise in uranium prices on the world market' and called for 'a rational price to facilitate the establishment of a stable nuclear fuel system worldwide and allow nuclear to compete with other energy sources.'[30] China would have to invest in uranium ventures in Africa, Kazakhstan, Australia and Canada to ensure 'greater security' of its uranium supplies.

In 2005, CNNC approached UraMin, a UK-registered emerging African uranium producer with mineral rights in Namibia, South Africa, Mozambique, Botswana, Chad and the Central African Republic, with the intention of buying uranium from or taking over UraMin. 'We are now facing a new era of uranium politics or rather "Uranium Politique",' cried the media. Competition for uranium resources in Africa was interpreted by politicians, scholars and the media as one piece of a larger picture of rivalry for resources and political influence in Africa. On several occasions, the officials from CNNC made it clear that China would not rely on any single supplier of uranium because of energy security considerations. It approached Namibia as well as Niger and the official visits made by Chinese high-level officials assisted and facilitated this pursuit of access to uranium in Africa. Rivalry for uranium resources around the world led to a nearly 600% increase in the world uranium price from the level of 2005 to US$351/kgU by mid-2007. In turn, rising prices triggered further fierce competition between CNNC and many other nuclear powers in negotiating with Namibia and Niger for their uranium resources.

In early 2007, as both CNNC and Russia were trying to get hold of UraMin, Areva came in and quickly concluded a deal of US$2.5 billion to gain control of a 100% share of UraMin. The deal allowed Areva to produce 7000tU by 2012 in South Africa, Namibia and the Central African Republic. After Areva took over UraMin, CNNC moved into negotiation with Areva to secure its uranium supplies from Africa in different forms. A deal of €8 billion was made at the end of 2007 for Areva to supply uranium, conversion, enrichment and fabrication services for up to two decades for two new 1700MW EPRs in Taishan, and Guangdong. Part of the deal on uranium was equivalent to 35% of the production of UraMin.[31]

In Niger, a former French colony, French companies had always had a dominant presence. China and Niger resumed their diplomatic relations in 1996 and leaders from both sides increased their official visits in the 2000s, as China had an eye on Niger's resources while Niger in turn wanted more assistance from China. In 2001, the presidents from both countries visited each other and a year later, the second meeting of bilateral joint economic and trade commission was held in China. As a result, trade between the two countries doubled from US$6.48 million in 2001 to US$14.7 million in 2002, all of which was made up of China's exports to Niger, mainly rice, textile and telecommunication materials. In addition, China also offered scholarships for students from Niger to study in China and sent medical teams as part of the general assistance to the country. With a rapidly developing diplomatic and trade relationship, the Chinese energy companies rapidly increased inroads into Niger. Two uranium companies attached to CNNC and CNEC started exploration projects in Niger, which led to a deal worth US$140 million. The agreement included the construction of a uranium mine with an annual output of 600 000 tonnes by Sino-Uranium, a coal-fired power plant and a hydro-metallurgy plant to be built by Sinohydro Corp. The deal was supported by the Export–Import Bank of China as part of a strategic agreement between this policy bank and the two companies. To prepare for the deal, a year earlier a joint venture had been created with CNNC holding 37.2% of the shares, the Niger government 33% and another Chinese company 24.8%.[32] In April 2009, China granted Niger another US$95 million concession loan from its Export–Import Bank to boost this uranium mining project, which would come online by 2010, with an annual output of about 700tU.[33]

China's efforts to get access to Niger's uranium have not been without difficulties. Behind its broad and often rhetorical promotion of being a responsible stakeholder in building a 'harmonious world' and promoting 'peaceful development', China had to deal with the reality that ultimately it went to Africa only to obtain access to the resources. As an increasing number of Chinese scholars were asking: 'What makes us different from the old imperial powers going to Africa to explore their resources?' Perhaps, one major difference was that China did not know Africa and African people as well as European powers and this lack of understanding was a major contributor to the difficulties Chinese companies encountered and to backlashes of the bilateral relationships between China and some African countries.

For example, after Sino-Uranium formed a joint venture with three local partners to explore and mine uranium in Niger, the project had to be shelved in 2007 because of local unrest. Chinese personnel were kidnapped by anti-government militants. Some Western observers in Beijing said that the Chinese government paid a ransom to free the captives from the rebels who were fighting against the Niger government. Tension arose between China's Ministry of Foreign Affairs and its large state-owned corporations that were in Africa to get access to its resources. When staff of these corporations found themselves 'in trouble', invariably they waited for foreign affairs officials to clean up the mess and fix the relationship.[34]

Sino-Uranium officials admitted the difficulties in its operation in Niger: 'In addition to bad weather, our people have to deal with diseases, poverty, ethnic conflicts and political instability', the deputy general manager of Sino-Uranium explained:

> Some 20 years ago, a Japanese team came to Niger, looking for natural resources, including uranium. They gave up when they reached the centre of the desert. It was just too difficult. 20 years later, we are here. It is a high-risk investment and it is a very difficult operation. I constantly worry about our people working there. But this is China's first uranium mine and we have to succeed.[35]

While the competition in Africa intensified, China moved into the Australian market. While most media and public attention in Australia focused on Chinese companies trying to take over iron ore companies, such as Fortescue, Chinese investors such as CITIC Australia and Sino-Uranium were targeting small 'grass-root' uranium exploration companies in South Australia. In the first part of 2007, South Australia saw a flood of new initial capital investment worth more than Aus$50 million (US$43 million) in its resource companies: such as CITIC Australia taking a 19.9% stake in tiny Marathon Resources in May 2007 and Aus$4 million for a similar stake in the IPO of Southern Uranium.[36]

Another targeted country is Kazakhstan, which holds 15% of the world's known uranium reserves. It was already the theatre of joint ventures with Russian companies, Cogema (France), Cameco (Canada, the UK) and Korea Hydro & Nuclear Power Corporations. In the immediate post-Cold War era, serious concerns were raised about the nuclear weapons left in Kazakhstan, a country that fortunately decided to disarm – 'a choice it reached due to a combination of international pressure, a desire to integrate into the international community, and assured Western assistance with dismantling its nuclear weapons and facilities.'[37] More than a decade later, Kazakhstan wanted to: (a) expand its uranium production; (b) become a significant supplier of nuclear fuel and (c) produce its own nuclear power. China saw these developments as a great opportunity to strengthen its 'cooperation' with Kazakhstan on a variety of issues, one of which was uranium supplies. As part of the efforts of Shanghai Cooperation Organisation, China and Kazakhstan established the Kazakh–Chinese Committee for Cooperation in 2004, which in the following years worked on several major pipeline projects, and cooperated on nuclear energy.

Until the early 2000s, only CNNC had a licence to purchase uranium or uranium products from overseas markets. For example, according to the former CEO of CGNPC, in 1999 Russia was selling enriched uranium converted from it several thousand nuclear warheads, and the price of uranium was as low as US$11/pound. Unfortunately, CGNPC missed the opportunity because it did not have a licence to purchase enriched uranium on the international market. With the expansion of China's nuclear energy program, the government extended licensing on uranium trade to include CGNPC. Now both CNNC and CGNPC can purchase enriched uranium from international markets.[38] While CNNC led an overseas expansion, the central government removed some restrictions to allow other Chinese companies to invest overseas to secure markets, technology and resources abroad. This coincided with many large Chinese state-owned corporations that had accumulated substantial economic wealth for their own disposal since the mid-1990s.

One main player in search for overseas uranium is CGNPC. In December 2006, Kazatomprom, a state-run energy company that oversees uranium production in Kazakhstan, and CGNPC signed the Strategic Agreement for a Mutually Beneficial Partnership, which led to the signing of a series of memoranda on cooperation between the governments of China and Kazakhstan in September 2007. Two memoranda signed by Kazatomprom and CGNPC defined the most important directions of strategic partnership in the field of energy, including the establishment of joint ventures for natural uranium production and Kazatomprom's investments into the nuclear power industry in China. The memoranda stipulated that all natural uranium produced by joint Kasakhstani-Chinese enterprises would be delivered to China in the form of 'high value added' nuclear fuel products. A year later, in October 2008, Kazatomprom and two Chinese nuclear companies, CNNC and CGNPC, signed another strategic partnership, which included long-term supplies of natural uranium for the nuclear power industry in China, development of uranium deposits on Kazakhstan territory (jointly with Chinese partners), fabrication of fuel for Chinese nuclear power plants and new lines of activity, that is, the construction of power stations in China.

Kazatomprom controls 51% and its Chinese partners control 49% of the stakes in three major uranium mines: Irkol with a total annual production capacity of 750 tonnes of U^{308}, Semizbay with a total annual production capacity of 500 tonnes of U^{308} and Zhalpak with a total annual production capacity of 750 tons of U^{308}. As Kazakhstan increased its uranium production from 6637 tonnes in 2007 to 9445 tonnes in 2008, it also increased its uranium export to China. In May 2009, the two sides agreed that the total exports of uranium to China would reach 24 200 tons by 2020.

By 2007, Sino-Uranium had its joint operation in Niger and exploration projects in Namibia. It was also negotiating with Kazakhstan, Algeria, Zimbabwe, Australia and Jordan.[39] It was clear that there were high risks involved in investing in uranium exploration and mining, especially in countries that are politically unstable. Moreover, it typically takes many years for a uranium mining project to move from discovery to production. This long process requires high finance and entails political risks. This means that despite the recent activities, it will be some time before China can expect to see any products.

Fuel production

Getting access to uranium is only one step towards securing nuclear fuel supplies for nuclear power plants and the price of uranium is only a minor part of the cost of reactor fuel for PWRs. Following the mining of uranium ore and the production of uranium ore concentration (U_3O_8, known as yellow cake), U_3O_8 is then converted to hexafluoride (UF_6), which is then enriched to increase the proportion of the U^{235} isotope from 0.71% in natural uranium to the level required for nuclear fuel (usually in the range of 3.5–5%). Enrichment is measured in separative work units (SWUs). Enriched UF_6 then has to be made into nuclear fuel rods (the process being called fuel fabrication) that are used to power the reactors (see Figure 7.1). If the uranium price is, for example, US\$130/kg, the price for nuclear fuel is about US\$1600/kg, which translates to about 0.5c/kwh. The price of nuclear fuel, meanwhile, varies from country to country and that made in China tends to be far more expensive.

Spent fuel from nuclear reactors has three components. Fission fragments that make up about 4% are intensely radioactive and need to be isolated for about 500 years until their radioactive level falls below a level of concern. Uranium makes up 95% and is negligibly radioactive. The remaining 1% consists of highly toxic substances, including plutonium. These are long-lived and have to be kept out of the biosphere for hundreds of thousands of years, or treated to somehow decrease the required isolation time. There have been debates on how to handle the long-lived substances. The US advocated the 'once through' fuel cycle, in which the spent fuel from PWRs was kept intact and disposed of untreated in a geological repository. Others, mainly the French, advocated reprocessing the spent fuel to separate the plutonium, blending it with uranium from the same spent fuel, and using this 'mixed oxide fuel' (MOX) in their LWRs (this is part of the back-end of

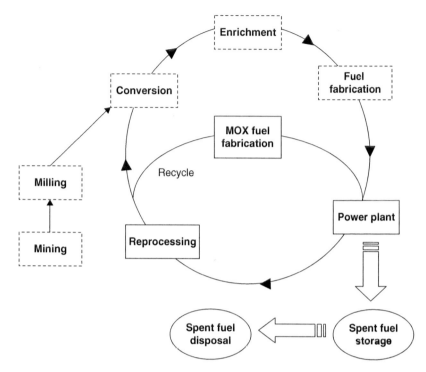

Figure 7.1   Nuclear fuel cycle

the nuclear fuel cycle). This can increase the energy for a given amount of enriched uranium fuel by about 30%.

Enrichment technology is extremely sensitive and strategic and is under close government control and supervision, even where carried out by private-sector corporations. The spread of enrichment technology to countries that do not yet possess it is a matter of great concern and subject to the control of the IAEA and other multilateral agreements. The recent interest in nuclear energy worldwide has raised serious concerns about nuclear proliferation not because of electricity generation from NPPs, but because of the potential spread of enrichment and reprocessing technology.

China is one of the nine countries in the world that have developed enrichment capacity, but in all stages of nuclear fuel production its capacity is limited. By the end of 2007, China had a UF_6 conversion capacity of about 3000t/U, roughly 1.2% of the world's total capacity. Its enrichment capacity was about 1000×10^3 SWU, 1.3% of the world's total.[40] The capacity is far short of meeting the domestic demand. By comparison, CNNC is just not in the same league as the big producers.[41]

China started a programme of producing fissionable materials in the late 1950s and its 'initial decision in favour of uranium over plutonium came in April 1960 [when] Beijing decided to concentrate on the production of enriched uranium in Lanzhou.'[42] No weapons programme could have been possible if the country could not produce fuel. Despite its promise, the Soviet Union had no intention of helping China develop enrichment capacity for the same reason that currently the international community has strict regulations on which country can develop nuclear fuel capacities. After the Soviet Union withdrew its experts and materials from China, the Chinese had no choice but to develop their own technology for nuclear conversion and enrichment. It managed to build a small facility that adopted the gaseous diffusion technology. The facility was able to produce enough low enriched uranium 'to satisfy the needs for the development of scientific research and national defence.'[43] It was far short of meeting the demand when a nuclear energy programme was in place.

How to expand the capacity of nuclear fuel production and service was subject to debate too. Some wanted to expand China's existing enrichment capacity and devote more resources to develop the back-end fuel cycle. Others argued that most international NPPs vendors can also provide fuel services and it was possible to purchase these services. By the 1980s, China had already obtained technology of the front-end of the nuclear fuel cycle – from mining and milling, to conversion, enrichment and fuel fabrication. It did not have to demonstrate that it could do what other nuclear power states were able to do. The issue was how to secure the resources to expand its nuclear fuel capacity when both the revenue of the central government and the budget for defence-military was down.

To ensure the nuclear fuel supplies for Qinshan I, in the 1980s, the 2nd Ministry planned to expand its enrichment facilities in Chengdu in Sichuan province, the site of important civilian and military nuclear activities. Little progress was made because of the budget constraints and structural changes of the industry. In 1989, China applied for IAEA Safeguards so that it could import more advanced nuclear technology because its own centrifuge development programme went nowhere. Right after, negotiations took place with the Soviets and then Russians to import a gas centrifuge enrichment plant with a production capacity of about 500 kilograms of low-enriched uranium (LEU) each year. This centrifuge enrichment plant in Chengdu was expected to be able to 'produce about 200 000SWU per year, for Qinshan I.'[44]

Russia had its own reason for selling the technology to China after the Cold War ended. A director of nuclear material production at the Russian Ministry of Atomic Energy (Minatom) said that the fate of the sales project to China would 'depend on whether or not we are kept out of uranium and other nuclear markets' in the US and elsewhere. If our expertise and capabilities are not rewarded in these markets, we will have to export our equipment and material elsewhere.'[45] For China, the central government had decided that nuclear fuel for Qinshan would come from domestic sources and foreign nuclear fuel would not exceed 30% for other nuclear power

Table 7.6 Forecast of SWU capacity requirement in China

	2000	2005	2010	2015	2020	2030
10^3 SWU	317.5	972.5	2078.3	4189	7413.8	10,213.2

Source: 李冠兴, '我国核燃料循环产业面临的挑战和机遇', (Li Guanxing, Challenges and Opportunities of Uranium Industry in China'), Uranium Geology, 24:5, 2008, p.262.

plants. The rationale behind this was that China had to become independent in supplying crucial energy sources.

In 1992, a bilateral nuclear cooperation agreement was signed: China agreed that centrifuges for its plant would be manufactured in Russia and then be installed at the Chinese site by Russian experts. In return, Minatom officials promised Russia would build the facility for CNNC based on its 'latest-design, subcritical centrifuges operation in Siberian locations', but would not transfer the technology to China.[46] This new enrichment plant with the centrifuge technology would be financed by the Chinese and was placed under IAEA safeguard provisions. This was a deal for mutual convenience: China was once again isolated and various degrees of embargo were imposed after the Tiananmen incident in 1989 while Yeltsin needed the business when Russia's economy collapsed and its military and defence segments were becoming increasingly distraught over the loss of the Soviet empire and their privileged position.

In the end, it was political considerations that dominated the 1992 bilateral agreement, but the Chinese negotiators failed to make sure that the agreement was fulfilled. Instead of selling China the most advanced technology, the Russian centrifuge facilities at Chengdu represented 'early Soviet-design centrifuge technology.'[47] During the construction, IAEA safeguard officials visited both sides to obtain design information and set up an inspection routine for the facility in Chengdu. The plant started its enrichment operation in 1997.

Meanwhile, Chinese scientists and engineers were trying to develop their own version of centrifuges at the same location. CNNC also started constructing 2×2 projects – two facilities for isotope separation and two facilities for fabrication. The two fabrication facilities where enriched uranium is fabricated into fuel rods to be fed into nuclear power reactors are located in:

- Yibin, Sichuan province – China's first fabrication facility was designed to play an important role in Qinshan I. Yibin Fuel Plant, now Jianzhong (建中) Nuclear Fuel Co. Ltd supplied and manufactured the first assemblies for the first load of Qinshan I with 11 tonnes a year of fuel assemblies. A second production line was built with technology transfer as part of the contract with Russia signed in July 1996 and the plant started its operation in 2001 to supply 26 tonnes a year of fuel assemblies to the Daya Bay units. The company has 5800 employees and about 30% of them are technicians with 229 high rank professionals. It increased its production capacity from about 150tU/y in 2000, to 200tU/y in 2006 and 400tU/y in 2008.[48]

- Baotou, Inner Mongolia (1998) – The operation was initially known as Baotou Nuclear Fuel Plant and is now called China North Nuclear Fuel Co. Ltd. It fabricates UO_2 fuel for two Canadian PHWRs in China at Qinshan III, after the Canadians supplied the first load of fuel. It also makes fuels for the HTR-10. It is also reported that the Baotou plant will eventually manufacture fuel for AP1000, which uses 4.95% of uranium, higher than the normal enriched 3.2%.[49]

Lanzhou, Gansu province and Chengdu, Sichuan province host two enrichment facilities that initially had gas diffusion technology and later introduced gas centrifuge technology to produce low-enriched uranium (LEU). They supplied the first fill and replacement for Qinshan I and Daya Bay. According to IAEA, CNNC is developing its enrichment plant in cooperation with Russia's Atomenergoprom and may increase its enrichment capacity to supply most or all of its growing domestic requirements. It is expected that the Chinese enrichment capacity will increase from the current 1.7% of the world's total to 5.5% by 2015. Currently China, with its enrichment capacity of one million SWU, can meet its domestic demand of enriched uranium. However, as the country expands its nuclear generation capacity, even if all its fuel enrichment capabilities were for civilian applications, domestic production would still be insufficient. The portion of foreign supplied nuclear fuel will rise to 50%.

Back-end of the nuclear fuel cycle

The back-end of the nuclear fuel cycle refers to the stage where spent fuel from a NPP is reprocessed and then used in combination of uranium to generate electricity. In countries where there is limited nuclear power generation capacity, spent fuel is often placed in on-site interim storage facilities until it is eventually removed to a permanent disposal facility. In some countries, spent fuel is reprocessed into plutonium that can be fed back into the enrichment and fuel fabrication process and used to fuel power plants.

The main purpose of reprocessing is to better utilise natural resources by recycling the remaining uranium and plutonium, thus reducing demands for fresh uranium mining and milling, and ensuring a more sustainable and long-term use of nuclear energy. Reprocessing also reduces the size of radioactive waste. Building reprocessing capacity is expensive and reprocessing produces plutonium that can be used in the production of nuclear weapons. The proliferation and economic concerns have been the main reasons that some countries have not developed, or have abandoned plans to build reprocessing facilities, as the US did in 1970. Currently, four countries – France, Japan, Russia and the UK – have reprocessing facilities, but only France and the UK operate commercial operations.

The Chinese nuclear community has long argued that the country should build an integrated nuclear fuel cycle industry. Developing reprocessing capacity has come to the foreground in recent years when the government decided to expand the country's nuclear power plants. Two arguments were presented to support this programme: (a) efficient utilisation of uranium; and (b) reducing the burden of accumulation of nuclear waste. Three unspoken reasons were behind the push to build an integrated nuclear fuel cycle by those at CNNC: (a) China should have control of the most advanced technology for civilian as well as military purposes; (b) the industry needed the government's continuing support and (c) it is prestigious to have the ability to build an integrated nuclear fuel cycle.

Throughout the 1980s and 1990s, the industry was under pressure to cut its programmes, lay off employees and make ends meet. Building reprocessing capacities, though expensive, was seen as a way to save the team because the central government might support the programme based on the principle that it would support R&D in 'high technology'. Few argued against building reprocessing capacities because of the military applications and political implications. R&D needed for this project, however, has been delayed year after year because of an array of economic, technical and logistical problems. The internal bickering within the nuclear sector, and between the nuclear and other energy sectors, and among departments and ministries of the central government, has been the major obstacle to addressing the issue in a coordinated manner.

In 1995, CNNC said that it planned to construct and operate a commercial-scale reprocessing plant by 2010. When building the reprocessing capacity was placed on the agenda of the central government in the late 1990s, it caused international safety and proliferation concerns. Officials from US agencies acknowledged that it was not clear why China would opt to reprocess spent fuel, but they suspected that 'there is an influential group that wants to build reprocessing plants'[50] and urged China not to proceed with the plan.

There was opposition in China too. It was argued that a closed nuclear fuel cycle was a complicated system; it would take a long time and a huge amount of investment to build. Its requirement for resources was in direct competition with other more urgent needs in the industry (at the time when so many of its workers were laid off). There were also inherent security concerns involved in developing reprocessing technology and fast breeder reactors. The possible use of nuclear materials and nuclear technologies for building nuclear weapons (proliferation resistance), and the spread of the sensitive technologies of the nuclear fuel cycle to where they might endanger the peace and stability of the world, would bring China in direct conflict with other countries at a time when it could not afford to do so. Finally, it would not be cost-effective or even necessary to develop the indigenous closed fuel cycle industry when its nuclear energy industry was not even developed. Given the domestic and international oppositions, the government invested limited resources in the programme.

Since the early 2000s, renewed efforts have been made as the country's nuclear energy programme expanded. For scientists at the CIAE, developing reprocessing capacity is only a step forward towards making nuclear energy sustainable and renewable. If PWR is the technology today, fast reactors would have to be the technology for tomorrow, not only to reprocess spent fuel to reduce the size of radioactive waste but also to produce more fuel than they consume. With fast breeder reactors, the utilisation rate of uranium would increase from 1% to 60–70%. The total world uranium reserves could last 30 times longer than if PWR open cycle technology was used (see Table 7.7). Chinese scientists want the technology and insist that it is essential for them to develop it because of the limited domestic uranium reserves. At the current consumption rate, uranium reserves in the world will last only 60–100 years with a light water reactor open cycle, shorter than coal reserves (155 years).[51] This is not and should not be a major concern today, and especially for those countries that will not have a large-scale nuclear power programme, but many believe it to be a key concern for Chinese nuclear development.[52]

According to senior scientists at CIAE, a pilot reprocessing plant with a capacity of 50tHM/a has been built and hot testing was underway in 2009 for the stable operation of the plant. China is planning to build a commercial reprocessing plant with a capacity of 800tHM/a by 2025. This plant will be constructed through international cooperation. CIAE has been undertaking research on a fast breeder reactor for some time. In 2007, the central government deliberately slowed the process of commissioning the plant for safety and security reasons. In 2009, the policy changed when the National

Table 7.7   Lifetime of energy resources (years of present annual consumption rates)

Coal	Gas	Uranium with a light water reactor open cycle	Uranium with fast breeder reactors
155	65	100[a]	> 3000

Note: [a]The Chinese estimate of the lifetime of uranium supplies with LWR was 60 years, while IAEA and NEA estimated 100 years.

Source: NEA, *Nuclear Energy Outlook 2008*. Paris: OECD, 2008, p.199.

Energy Administration (NEA) listed the project as a step towards 'independent nuclear energy technology development'.

In January 2010, the NEA announced 16 energy R&D centres and the CIAE's fast reactor is one of them. Scientists at CIAE see the fast breeder reactor as the means leading to sustainable nuclear development; it is designed to recycle the separated plutonium from the reprocessing process. If the fast reactor is not in place by 2025, the reprocessed plutonium can still be used with the remaining uranium as MOX to feed into the nuclear power plants. In sum, for now, the back-end of the nuclear fuel cycle in China is still at the testing stage, while R&D is being carried out as the technical support to the stable operation of the pilot repressing plant.[53]

Another reason for the industry to push for the reprocessing and fast reactor technologies is the accumulated spent fuels from the existing reactors. Radioactive waste might yet become a public concern in China. It has not been a problem so far because of the limited number of nuclear power. However, radioactive waste in China has accumulated at a fast rate (as shown in Table 7.8).

Accumulation of spent fuel raises serious concerns about waste management.[54] Each NPP has its own on-site storage facility and specific treatment facility, generally including waste separation, cementation and compaction. For the Daya Bay project, two near-surface facilities have been built. One is the Beilong repository in Guangdong province, about 5 km from the Daya Bay nuclear power station. The design capacity of Phase 1 is 80 000 cubic metres, and the capacity for the initial phase is 8800 cubic metres. The SEPA approved its assessment report of environmental impact in March 1998.

Another major repository is under construction near the Gobi Desert in Gansu province. This will have a 200 000 cubic metre capacity when it is finished. By the early 2000s, it was about 20 000 cubic metres, including six disposal units. It has already accepted some waste from Daya Bay because normally spent fuel is kept on-site until the end of the reactor life – between 40 and 60 years. The storage facility at Daya Bay, however, was full only 10 years after it started operation in 1994. The two French-designed reactors at Daya Bay were equipped with a small spent fuel storage area similar to that built in France, where small storage capacity is not a problem because spent

Table 7.8   Forecasts of spent fuels in China

	2000	2005	2010	2015	2020
tHM/y	54.4	166.8	272.8	537.9	983.2
Accumulated (10^4 tHM)	348.2	885.6	1954.8	4055.7	8011.8

Note: tHM – tones of heavy metal.

Source: 李冠兴, '我国核燃料循环产业面临的挑战和机遇', Uranium Geology, 24:5, 2008, p.264.

fuel is routinely shipped to Cogema's reprocessing facilities at La Hague. In China, however, no such solution is yet available.. Currently, CGNPC, the operator of the Daya Bay reactors, is paying CNNC to remove its spent fuel from Daya Bay and transport it in casks to Gansu province where CNNC is building a reprocessing facility. A permanent storage place is also under review in Gansu province.

The most advanced model of nuclear reactor, AP1000, would generate about 35 tonnes of intermediate level and low level waste each year. China will build two or three units a year in the coming decades. How to manage the rapid accumulation of toxic radioactive waste is becoming one of the most pressing challenges in China. This has become one of the main motives for Chinese scientists to develop an integrated fuel cycle industry so that the size of the waste would reduce significantly.[55]

Meanwhile, international cooperation is taking place to deal with the twin challenges of efficient utilisation of uranium and managing radioactive waste. In 2007, as a side agreement for China to purchase two EPRs from Areva, Areva and CNNC signed an agreement to study the feasibility of cooperating in constructing a reprocessing and plutonium complex, as a way of 'industrial cooperation in the back-end of the nuclear fuel cycle.' Soon after Areva and China agreed on cooperation, Russia approached China for a similar bilateral collaboration 'with the goal of setting up a commercial reprocessing complex in China' as an extension of their decade-long partnership in centrifuge uranium enrichment.[56] In late 2008, AECL signed an agreement with Chinese authorities to develop technology to recover uranium from the spent fuel of China's light water reactors so that it could be used in the two 728-MW Qinshan CANDU reactors. Canada currently does not have the reprocessing capacity, and neither does China.

Conclusion

Nuclear fuel is the main concern for those calling for slow nuclear energy development.[57] China has limited natural uranium reserves and its nuclear energy programme depends on resources imported or developed by Chinese firms in overseas mining. This is the strategy adopted by the Chinese government– expanding domestic uranium exploration, mining and milling, encouraging Chinese firms to invest in overseas mines, and importing from the major producers. Together they are known as 'two sources, two markets' (两种资源, 两个市场). Entering the global uranium industry late that is dominated by a few multinational conglomerates, Chinese companies are investing in uranium exploration and mining in some very unstable and highly risky countries.

There is a global market where nuclear power plant operators can buy nuclear fuel services under the strict supervision and monitoring of IAEA. While China supports the efforts of the international community to adopt a multilateral approach to guarantee nuclear fuel supplies to those countries wishing to develop and expand their nuclear generation capacity, it wants to be the supplier of nuclear fuel rather than depending on the external supplies.

Finally, it has been a strong desire of the Chinese nuclear community to develop the back-end of the nuclear fuel cycle, specifically the spent fuel reprocessing and fast breeder reactors. This raises two issues: one is that given that the technology used in enrichment and reprocessing can be used to make weapons, China would have to put in place better and more effective regulations. The other is that developing back-end capacity is difficult, expensive and time consuming. CNNC is the only player in the field and many of its scientists treat nuclear fuel production as a scientific development issue rather than a commercial one. The industry thus faces the challenge of how to balance the immediate need for fuel and the future sustainable nuclear development.

Notes

1. Mohamed ElBaradei, 'Speech at Beijing International Ministerial Conference on Nuclear Energy in the 21st Century', 20 April 2009, p.4.
2. Price, 1990, p.27.
3. NEA/IAEA, 2008, p.20.
4. NEA, 2006, p.179.
5. NEA, 2008, p.157.
6. NEA/IAEA, 2008, p.154.
7. Ibid, p.154; Mark Hibbs, 'CNNC Uranium Prospecting Operation May be Shunted to Beijing Ministry', *Nuclear Fuel*, 24:9, 3 May 1999, p.5.
8. '核地质勘查队伍管理体制改革实施方案', ('Restructuring Plan of Uranium Industry'), 22 June 2006, http://www.hgydzj.jx.cn.2006-6/2006622154824.htm.
9. For the reform of SOEs and laying off state employees, see Steinfeld, 1998.
10. 孙勤, 核地质系统结构调整的初步设想, 中国核工业, (Sun Qin, 'Preliminary Plan of Restructuring Uranium Industry', *China Nuclear Industry*), no.2, 1998, pp.23–25.
11. 戴民主, '江西省核电发展与铀资源保障', 江西能源, (Dai Minzhu, 'Nuclear Development and Uranium Supplies in Jiangxi', *Jiangxi Energy*), 4, 2008, pp.1–7.
12. NEA, 2008, p.157.
13. NEA/IAEA, 2008, p.29.
14. 叶柏庄, '我国铀地勘可持续发展的战略思考', (Ye Baizhuang, 'Sustainable Development of China's Uranium Resources', *Advances in Earth Science*), 21:11, 2006, pp.1134–39.
15. NEA/IAEA, *Uranium 2007: Resources, Production, and Demand.* Paris: OECD, 2008, p.31.
16. 郑大瑜, '跨入新世纪迎接新挑战', (Zheng Dayu, 'Challenges in the New Millennium'), *Uranium Geology*, 17:1, 2001, pp.1–4; Y. Chen, 'The Recent Progress of Uranium Exploration in China', in IAEA, 'Uranium Production and Raw Materials for the Nuclear Fuel Cycle Supply and Demand, Economics, the Environment and Energy Security', Proceeding SeriesIAEA-CN-128, May 2006, pp.57–62.
17. NEA/IAEA, 2008, p.154.
18. Ibid.

19. 肖新建 高世宪, '我国核能资源保障研究', 国家发展和改革委员会宏观经济研究院, http://www.amr.gov.cn/qikanshow.asp?articleid=191&cataid=5, accessed on 1 March 2010. (Xiao Xinjian and Gao Shixina, 'Analysis of Uranium Reserves in China', Macroeconomic Research Institute of the State Council).

20. 国土资源部与国防科工委, ' 关于加强铀矿勘查工作的若干意见', (国土资发 [2 0 0 8] 4 5号), 17 April 2008. (Ministry of Land and Natural Resources and COSTIND, 'Suggestions on Uranium Exploration').

21. Nuclear Energy Agency and IAEA, *Nuclear Energy Outlook 2008*, Paris: OECD, 2008, p.5.

22. Hong and Sun, 2006, p.620.

23. Jiang Zemin, 'Accelerating Reform and Open Up', *Beijing Review*, 26 October 1992, pp.9–32.

24. Nolan, 2001.

25. Harding, 2006, p.57.

26. Downs, 2000, p.15.

27. Ibid, p.20.

28. Friedberg, 2006, p.23.

29. Quoted from Mark Habbs, 'CNNC: China Will Satisfy U Demand through Combination of Sources', *Nuclear Fuel*, 33:12, 16 June 2008, p.15.

30. Ann MacLachlan, 'As China Worries about Uranium Cost, CNNC Executive Seeks "Rational Price"', *Nuclear Fuel*, 30:20, 2005, p.10.

31. Ann MacLachlan, 'Areva, China Seal Long-term Mining, Fuel Cycle Pact', *Nuclear Fuel*, 32:25, 3 December 2007, p.1.

32. 'Chinese Firms sign $140 million Uranium Project in Niger', Xinhua, 9 April 2008, and Andrew McGregor, 'Mining for Energy: China's Relations with Niger', China Brief, 7:18, 3 October 2007.

33. 'China Extends Niger $95m Loan for Uranium Project', *Router*, 24 April 2009.

34. *China Quarterly* in 2009 published a special issue on 'China and Africa: Emerging Patterns in Globalisation and Development', no.199. Contributors to this special issue have raised a set of questions about globalisation and development, and how China's entry to Africa may or may not differ from the previous attempts of the 'scramble of Africa'. Uranium is only one of the resources China has been targeting in Africa.

35. 黄卉, 海外找铀, 我们阔步向前, 中国核工业, (Huang Hui, 'Investing in Overseas Uranium', *China Nuclear Industry*), no.2, 2008, pp.36–37.

36. 'Natural – China Stokes Furnace in Iron Ore War', *Asianmoney*, 28 December 2007.

37. Togzhan Kassenova, 'Kazakhstan's Nuclear Ambitions', *Bulletin of the Atomic Scientists*, 28 April 2008.

38. 冯志卿, '中广核以核养核', 中国投资, (Feng Zhiqing, 'CGNPC's Rolling Development', *China Investment*), March 2008, pp.66–70.

39. 邱建刚, '抓住铀地矿行业大发展历史机遇', (Qiu Jiangang, 'Grabbing the Opportunity of Uranium Development'), *Uranium Geology*, 24:4, 2008, pp.193–94.

40. '探密中国核电 "粮仓"', 11 August 2009, http://www.cnnc.com. (Searching for China's Uranium').

41. NEA, 2008; NEA/IAEA, 2008.

42. Lewis and Xue, 1988, p.113.

43. Ibid, p.112.

44. Mark Hibbs, 'China's Centrifuge SWU Plant Up and Running, Minatom Says', *Nuclear Fuel*, 22:2, 27 January 1997, p.3; 钱福源, 重振雄风, 再创辉煌, 中国核工业.

45. Mark Hibbs, 'Enrichment', *Nuclear Fuel*, 17:22, 26 October 1992, p.4.

46. Mark Hibbs, 'Russia to Build Centrifuge SWU Plant at China's Shenzen Technology Centre', *Nuclear Fuel*, 18:21, 11 October 1993, p.1.
47. Mark Hibbs, 'China's Centrifuge SWY Plant Up and Running', *Nuclear Fuel*, 22:2, 27 January 1991, p.3.
48. 探秘中国核电'粮仓', 12 October 2009, http://bbs.cnpjob.com/thread-36805-1-1.html.
49. Mark Hibbs, 'CNNC to Boost Production Capacity at Two Fuel Fabrication Plants', *Nuclear Fuel*, 33:7, 7 April 2008, p.3.
50. Mark Hibbs, 'Chinese Separation Plant to Reprocess Spent Fuel HEU Fuel', *Nuclear Fuel*, 22:1, 13 January 1997, p.3.
51. 顾忠茂 黄齐陶，我国亟需尽快启动快堆核能系统的技术开发，中国能源，27:4, April 2005, pp.9–11; 顾忠茂 王乃彦，我国核裂变能可持续战略研究，中国能源，27:11, November 2005, pp.5–10; NEA, *Nuclear Energy Outlook 2008*, Paris: OECD, 2008, p.199.
52. 李冠兴, '我国核燃料循环产业面临的挑战和机遇', *Uranium Geology*, 24:5, 2008, pp.257–67.
53. GU Zhongmao, 'Securing Nuclear Fuel Cycle When Embracing Global Nuclear Renaissance', speech at the International Ministerial Conference on Nuclear Energy in the 21st Century, 20–22 April, 2009, Beijing, China.
54. Forum for Nuclear Cooperation in Asia, 'FNCA Consolidated Report (China), Updated as of March 2007', 22; Forum for Nuclear Cooperation in Asia, 'The Consolidated Report on Radioactive Waste Management in FNCA Countries', March 2003.
55. Gu Zhongmao, 'Securing Nuclear Fuel Cycle When Embracing Global Nuclear Renaissance', speech at the International Ministerial Conference on Nuclear Energy in the 21st Century, 20–22 April 2009, Beijing, China.
56. Mark Hibbs, 'Sino-Russian Fuel Cycle Cooperation Poised to Move Beyond Enrichment', *Nuclear Fuel*, 34:9, 4 May 2009, p.1; Ann MacLachlan, 'Areva, China Seal Long-term Mining, Fuel Cycle Pact', *Nuclear Fuel*, 32:25, 3 December 2007, p.1.
57. Wang, 2009.

8
Who Cares? The Public and the Environment

Worldwide, nuclear energy development has been shaped by public opinion, which has to be disaggregated into three levels – elite, sub-elite and popular.[1] In some societies, elite opinion holds more sway than the other two groups, while in others the popular views can have a significant impact on decision making. The balance also changes with time and across issues. How public opinion shapes nuclear policies depends on which aspects of nuclear energy development are debated: issues of environment, safety, siting, waste management or decommissioning.

Environmental pollution was one the issues that the Chinese government decided to tackle as early as 1979 when the economic reform had just begun. Energy production and consumption was identified as the major contributor to environmental pollution. The central government invited American scientists to China in 1978 and 1979 to discuss the possibility of developing renewable energy, solar, wind, geothermal and tidal wave.[2] In the following three decades, environmental pollution has been a perennial challenge for policy makers, but what constitutes environmental pollution was constantly changing. In the early years, direct coal-burning was identified as the main threat to environment and as a major reason for energy inefficiency. Nuclear energy was advocated as a clean, safe and cheap energy to replace the dirty coal to deal with both air and water pollution and to meet rising demands for modern energy.

Nuclear energy development did not raise environmental concerns in China until recently mainly because the desire to meet rising energy demands overwhelmed concerns of pollution. Nuclear energy involves two types of environmental concerns: one is 'offensive' and the other 'defensive'. At every stage, nuclear power production, from uranium mining (such as radioactive tailing) to electricity generation (spent fuel) to decommissioning, produces some adverse impacts on the natural environment. Scientists have been developing technologies to limit the impact of these offensive environmental pollutions, which are hardly noticed or talked about by the general public. Defensive pollution refers to the way in which the public and politicians try to protect what they have – people, property, scenery or the sense of safety and security. These threats to the environment may only be 'potential' and they may affect, or at least are perceived to affect, people living in the areas where nuclear power stations are built. The public responds to these potential negative environmental consequences of nuclear energy development, such as radioactive leakage into the air, land and water, or visual pollution, much more strongly and vocally than they react to the offensive ones.

At the core of this different reaction to offensive and defensive environmental consequences is the issue of 'dispersed benefits vs. concentrated costs'.[3] When people can benefit thousands of miles away and gain access to electricity without suffering the air and water pollution from burning coal to generate it, they support nuclear power, but not with any great enthusiasm that would make them active advocates of NPPs. For those living in the vicinity, the change of their livelihood, destruction of scenery and potential radioactive leakage are more than they want to tolerate; and the impact of these potential threats would be much more immediate. So they care and make a noise. The government has to reassure the public about the safety and security of NPPs with adequate and effective regulation in place. It is meanwhile often unwilling to push through a project when facing strong public opposition, even in China, because it is not worthwhile politically.

As economic reform has progressed, the policy-making process has become more open and organised interests (the sub-elite) have gained significant status by making their views known. The civil society and the 'right-define' movement, 'stemming from the substantial expansion of "public space" available for addressing issues of civil concern,'[4] are making inroads in shaping public policies. Indeed, a major NPP project in Shandong was put on hold largely because of the complaints and organised protests of civil societies.

This chapter examines how a series of issues have been balanced in the process of nuclear energy development: such as meeting the immediate need for modern energy, long-term energy security, the immediate human suffering from air or water pollution, and the long-term safety of radioactive waste. Since public policy-making is all about making choices, there is a dynamic process in which immediate and future problems and the costs of dealing with these problems are balanced.

Energy demands vs. environmental protection

Environmental protection and sustainable development have been proclaimed among the 'major tasks and important targets' in the FYPs since the 1970s, yet they were overwhelmed by the desire to meet rising energy demands. In policy terms, expanding electricity capacity was given the priority for two main reasons: (a) economic growth required stable supply of modern energy, and (b) air and water pollution from direct coal-burning had more immediate impacts on society that the government had to deal with. Policies adopted to reduce one type of pollution (dust and particles) by expanding thermal generation capacities then created serious problems of another type of environmental pollution – greenhouse gas emissions. Nuclear energy programme was designed to address both rising energy demand and climate change, but neither proponents nor opponents of nuclear energy made much difference in this development.

As discussed in Chapter 2, in the early stage of reforms, power shortages swept the country. In 1978, for example, the Shanghai government issued a document, stipulating the principles of power rationing. The city government allocated power consumption to enterprises, government agencies, schools, shops and the military. Those that consumed more than their quotas would have to pay heavy fines. Enterprises would have to rotate their shifts in order to avoid peak-hour usage of electricity. In 1979, Guangdong could supply only about 60% of the electricity that was required. In December 1980, the Chinese premier, Zhao Ziyang, admitted that the electricity shortage was the main reason that China's industry ran at about 70% capacity. In 1983, Beijing city government issued a similar document, requiring all enterprises to rotate their shifts and work on different days of the week to avoid peak-hour electricity demand. The price of electricity for peak-hour and for beyond-the-quota consumption was more than double the normal power tariffs.[5]

Meanwhile, environmental pollution was a grave problem throughout the 1980s and 1990s. In the early 1980s, the amount of smog and pollutants released into the atmosphere in China was about 20 million tonnes a year, which was twice the world average. The concentration of particulates of sulphur dioxide (SO_2) and nitrogen oxide (NO_x) in the air in the centre of Beijing exceeded the national standard by two to four times, and in Beijing the average dust (total suspended particulate) levels were about seven times greater than the US air quality standard. In rural areas, biomass fuel uses caused serious hillside soil erosion, excessive water runoff, deforestation and declines in soil fertility.[6] 'To prevent further increases in already unacceptably high levels of urban air pollution, as well as to economise on fuel, [required] the replacement of decentralised and uncontrolled burning of coal in households and enterprises with centralised, large-scale, environmentally controlled combustion to produce cleaner forms of energy (gas, electricity, steam, hot water) for distribution to final users.'[7]

This was what China did in the following two decades: a large number of coal-fired power stations, including many small and dirty ones, were built in the 1980s to replace direct coal-burning. China significantly increased its generation capacity, from 66GW in 1980 to 217GW in 1995. The proportion of coal used for power generation expanded from 20.13% in 1980 to 33.26%

in 1995 and to 45.64% in 1999. The efforts to reduce one type of pollution were more than balanced out by an increase in GHG emissions as the result of a rapid expansion of coal-fired thermal power stations and coal gasification, both of which were encouraged by the experts from the World Bank and UN Environment Programme (UNEP).

Nuclear scientists pushed for nuclear energy development for the same reasons. 'Environmental pollution from nuclear power stations [was] much less than power stations burning coal,' explained the president of the Chinese Nuclear Society in 1984. They released no sulphur dioxide, no nitrogen oxide, no dust or other dangerous substances (see Table 8.1). Even the amount of radiation released by a nuclear power station was less than that of a coal-fired thermal power plant.

Nuclear scientists' argument that a heavy reliance on coal had serious consequences for the environment – it polluted air, dirtied water and created hazards – might be convincing. To many policy makers, however, building NPPs would take too much time and resources that the country did not have to deal with either electricity shortages or pollution.

A series of policies was adopted for environmental protection: in 1989, the central government introduced the Environmental Protection Law. The 8th FYP (1991–95) listed environmental protection among the 'major tasks and important targets for the following five to ten years.'[8] After the Earth Summit at Rio de Janeiro in 1992, China adopted its own version of 'Agenda 21'. In the 1990s, environmentalists and nuclear establishment took the same line: fossil-fuel consumption should be controlled. 'Many environmental specialists indicate that one of the major measures to improve the Chinese environmental outlook is the development of nuclear power stations to replace coal power stations,' an official from CNNC said at the 9th Pacific Basin Nuclear Conference. 'The practice of world nuclear power development demonstrates that nuclear power is a safe, clean and economic energy source, it not only reduces the pressure on railway transportation of coal, but also benefits environmental protection.'[9] This statement was a carbon-copy of the speech some top scientists gave at the end of the 1970s when China decided to adopt a

Table 8.1   Early estimates of environmental consequences

	Radiation exposure (mr/year)	Sulphur dioxide SO_2 (tonne/year)	Nitrogen oxide NO_x (tonne/year)	Dust and other substance (ton/year)
Thermal power station	4.75	46,000–127,500	26,250–30,000	3500
NPP	1.8	0	0	0

Source: Jiang Shengjie, 'Developing China's Nuclear Power Industry', *Beijing Review*, 18 June 1984, p.19.

nuclear energy programme. Some Western observers in China at the time made similar statements:

> Nuclear power is even more appealing to China's planners when they weigh the environmental consequences of other power sources. Compared to serious air pollution problems caused by hydrocarbon combustion and the potential for environmental degradation from new large-scale hydro-power dams, nuclear power appears considerably more benign... Each 1,000MW in new nuclear capacity that displaces a coal-fired power plant could reduce China's potential carbon emission by roughly three million tonnes per year. If China's nuclear plants account for 5 percent of total energy consumption by 2020, about 125 million tonnes in carbon emissions could be avoided, or the equivalent of France's current total annual carbon emissions.[10]

Neither government officials nor the public saw nuclear energy as a viable alternative to coal, which, many argued, would be the main source of energy for China in the foreseeable future. This was the position presented by the Chinese delegate at the Kyoto Conference in 1999.[11] Meanwhile, the discussions on environmental pollution in the context of the inhaled particulates shifted to those on climate change, especially GHG emissions. With a rapid expansion of thermal generation capacity, China had become the second-largest GHG emitter, behind the US, producing about 14% of the world's GHGs by the end of the 1990s. Scientists in China pointed out the aggregate of pollutant emissions was far beyond the environmental loading capacity.

Since energy was at the centre of environmental problems in China, it was widely agreed among the science and nuclear community that China should pursue nuclear energy to deal with both rising energy demand and environmental problems simultaneously. This was the conclusion drawn by a team of government officials who wrote the report on nuclear energy development in China in 1999.[12] Unfortunately, a year earlier, Zhu Rongji announced: 'Let's put nuclear power on hold right now and put our emphasis on other power sources.'[13] In 2000, Zhu repeated the message that nuclear energy would not be the priority; developing hydro projects in the western provinces would help to alleviate both poverty in the west and power shortages along the coast.

The growth rate of carbon emissions escalated after 2002 as more coal-fired generation plants were built to deal with resurged power shortages. The total generation capacity rose by 9.9% in 2003, 9.74% in 2004 and 16.91% in 2005, and CO_2 emission per capita in China jumped from 2.89 tonnes in 2003 to 3.65 tonnes in 2004 (26.3% increase), then increased by another 6.3% in 2005, 10% in 2006 and 7% in 2007.[14] 'Burning coal contributes to 90% of the national total sulphur dioxide (SO_2) emissions, about 70% of the national total dust, nitrogen oxide (NO_x) emissions and carbon dioxide

(CO_2) emissions.'[15] China became the world's largest emitter of SO_2 and the second-largest emitter of CO_2.

Environmental pollution has been a major factor impeding sustainable economic and social development. China hosts 16 of the most polluted cities on the planet; air pollution alone, primarily from coal-burning, is responsible for more than 300 000 premature deaths a year. According to the SEPA, more than 70% of the country's river systems are badly polluted, more than 300 million people do not have access to clean water and more than 400 million people in urban areas do not have clean air. The problem also affects China's neighbours when pollutants generated in China are carried by the wind and river systems. There is a growing consensus that energy policy makers cannot ignore these environmental impacts to the extent that they did in the past.

The economic costs of environmental pollution are high. According to the World Bank, the associated costs reached 6–8% of GDP in China.[16] The political costs are even higher. The devastating impact on the environment has become the focus of a growing number of local protests by disgruntled citizens. In 2005 alone, more than 50 000 disputes on violation of environmental regulations were reported to different levels of government.

When environmental degradation and the unsustainable management of natural resources became an obstacle to further economic development, the well-being of the population and potential political stability, addressing environmental problems, in particular pollution caused by coal-fired electricity, became a state priority. Scientists and policy makers in China have agreed that nuclear energy could be developed as an alternative to 'supplement', not 'replace', the most polluting and carbon-intensive base-load source of electricity generation: coal. Indeed, nuclear energy is 'the sole energy that can substitute fossil [fuels] in a centralised way and in a great

Table 8.2   Average CO_2 emissions by energy source (Kg CO_2/kWh)

Energy chain	Average CO_2 emission
Lignite	1.2
Hard coal	1.07
Oil	0.9
Natural gas (combined cycle)	0.4
Solar PV	0.06
Wind (offshore)	0.014
Wind (onshore)	0.014
Nuclear	0.008
Hydro	0.005

Source: Nuclear Energy Agency, *Nuclear Energy Outlook 2008*. Paris: OECD, (2008), p.122.

amount with commercial availability and economic competitiveness.'[17] Among the various fossil fuels, coal emits significantly more CO_2 per unit of energy produced than either oil or gas, 'accounting for 40% of global CO_2 emissions although it only has a 25% share of total primary energy supply.'[18] Compared with all primary energy sources, nuclear emits the least CO_2 other than hydro per kilowatt hour generation (see Table 8:2).

The government in China has now accepted that environmental concerns have to be an integral part of its energy policy-making decisions and that nuclear energy is an alternative. This change of position has also been facilitated by the changing pattern of major pollution emissions in China over the past decade. Dust and soot from direct coal-burning and industrial pollutants have been significantly reduced because of the shift to using electricity for industrial production or for generating heat. As power generation capacity expanded, so did carbon dioxide emission.

In 2005 and 2006, the State Council approved the Medium- and Long-Term Energy Development Plan, 2004–20 and the Medium- to Long-Term Plan for Nuclear Energy Development , 2005–20. Both documents raised the alarm about deteriorating environmental problems as the result of a heavy reliance on coal and called for a quick expansion of renewable energy. Indeed, the very first sentence of the preface of the plan for nuclear energy stated, 'since nuclear energy does not emit any GHG, active promotion of nuclear energy expansion is one important policy for China'. It pointed out that the increased use of nuclear energy to supplement some of the growth in thermal generation capacity would help improve the condition of the environment and alleviate some climate change concerns. It is clear that environmental protection is only one reason for nuclear energy expansion. Changing the energy mix for energy security is as important as environmental protection.[19]

In sum, in the early 2000s, as China's rate of growth of carbon emissions rose steeply, especially after 2002, and pollution problems became more serious than ever before, the Chinese leaders realised that China was among

Table 8.3 Major pollution emissions in China, 2000–08

	2000	2001	2002	2003	2004	2005	2006	2007	2008
SO_2 (Mt)	19.95	19.48	19.27	21.59	22.55	25.49	25.89	24.68	23.21
Dust (Mt)	11.65	10.70	10.13	10.48	10.95	11.82	10.89	9.87	9.02
Ind. dust (Mt)	10.92	9.91	9.41	10.21	9.05	9.11	8.08	7.71	6.71
CO_2 (Mt)	3016.9	3217	3497	4045.8	4732.3	5059.9	5606.5	6027.9	N/A
CO_2/Pop (t)	2.38	2.52	2.72	3.12	3.66	3.88	4.27	4.57	N/A

Source: data on SO_2, dust, and industrial dust are from the Ministry of Environmental Protection, *China Environmental Year Book*, various years. Data on CO_2 and CO_2 per capita are from the International Energy Agency, *CO2 Emissions from Fuel Combustion, 1971–2004*. Paris: OECD (2006) and IEA, *Key World Energy Statistics*, various years. Paris: OECD.

the countries most vulnerable to climate change. They adopted serious measures to control the country's rising carbon emissions, including the plan to expand its nuclear energy capacity quickly.

The Public and nuclear energy

If environmental protection is now a rationale for nuclear expansion, the nuclear industry also faces its own environmental challenges.

For most of the history of nuclear energy development, the 'offensive' environmental consequences of the nuclear energy industry were rarely discussed except among those in the industry and those affected one way or another. The main focus was on the 'defensive pollution' – how to protect people and property from potential damages of the nuclear industry. The anti-nuclear movement was organised around the theme of protecting people and property and only indirectly addressed the issue of protecting the environment. This human face of the nuclear industry means that how the public reacts often determines whether a nuclear energy programme can be launched, expanded, put on hold or cancelled completely. This is even the case in China.

In the 1980s and 1990s, nuclear energy development in China was the subject of debates among the elite and it seldom drew much public attention. Even then, the government could not brush away whatever concerns there might be. It had to justify nuclear energy projects to the public domestically and internationally. As the reform progressed, the middle class has become much more willing to speak up on nuclear energy development and on environmental pollution. The nuclear decision-making process was becoming increasingly subject to public reaction. The typical problems associated with diverse benefits and concentrated losses have been the main hurdles of some projects in China in the 2000s too. The focal points of contention are often between those who want sufficient and reliable access to electricity and those who are concerned about the potential risks of nuclear projects in their backyard.

The first challenge China faced regarding the public reaction to nuclear development was the mass demonstration that broke out in Hong Kong following the Chernobyl disaster. The nuclear power plant meltdown in Chernobyl in Ukraine on 26 April 1986 sparked off an unprecedented debate in Hong Kong. In the first month after the disaster, the debate 'was largely confined to safety issues such as the likelihood of a nuclear accident at Daya Bay and its possible effects on the health of the Hong Kong population, the feasibility of an evacuation plan and other contingency measures.'[20] The debate, organised by the newly formed Joint Conference for the Shelving of the Daya Bay Nuclear Plant, soon escalated into mass demonstrations in June and July of the same year. Its 30-odd constituents included students' organisations, labour unions and, in particular, the influential 30 000-member

Professional Teachers' Union to fight against the project. Emotions ran high over the issue. 'Not even the debate on Hong Kong's political structure after 1997 has stirred such emotions,' said an Anglican minister and spokesperson for the anti-nuclear movement in the territory.[21]

However, there was no consensus in the anti-nuclear camp on the reasons for their opposition. Some anti-nuclear protestors were not opposing nuclear power *per se*; it was just that they did not want a nuclear power station so close to Hong Kong. Daya Bay is only 50 kilometres from the centre of Hong Kong. Some critics did not 'doubt the international yardsticks of the ultra-modern facility' – the French-designed and French-built pressurised water reactors – but 'human errors can break out at any time and in any place' and in particular they 'had little or no confidence in Chinese management and operation expertise.'[22] Some simply argued that there would be no guarantee that a nuclear power plant was 100% fault-proof; 'the danger is always there, regardless of the type of safety measures we take,' argued a scientist.[23] Others argued that the Daya Bay site was an invitation for disaster because it sat on an earthquake fault line.

Despite different reasons for their opposition to the Daya Bay project, the anti-nuclear campaign gained momentum during the summer of 1986 in Hong Kong. It had the support from many developed countries as the anti-nuclear movement spread around the world. Environmentalists and energy experts joined the forces. By August 1986, the organisers of the protests against the Daya Bay project in Hong Kong had collected approximately one million signatures on a petition and delivered them to Beijing, in the hope of stopping the project. Fuelling the anti-nuclear debate was the question of whether the Chinese government in Beijing would honour its promise made two years earlier in the Sino–British Joint Declaration on Hong Kong that China would not change Hong Kong's political and economic system and would keep its promise of 'one country, two systems' for 50 years after the return of the British colony to China in 1997.

For the Legislative Council (Legco) of Hong Kong, it was an issue of accountability because just one year earlier the Council was revamped, with two-fifths of its members directly elected. 'Once considered a rubber stamp for the more powerful policy-making Executive Council, Legco had become bitterly divided over Daya Bay.'[24] Some newly elected members at Legco demanded an open debate on the project, while others asked for a reassessment of the project by Britain's Atomic Energy Authority, which had already provided an assessment in mid-1985. In September, when the Chinese concluded negotiations with the British and the French on the project, Legco, after a four-hour, closed-door meeting, rejected a call by some members for a public debate on the project. 'One lasting legacy of the Daya Bay debate could be a growing conviction in Hong Kong that political mobilisation [would be] the only way to obtain concessions from Beijing,'[25] the media commented.

When mass demonstrations broke out in Hong Kong, officials of the Guangdong government flew to Beijing, asking for help and instruction. A minute of the meeting was eventually approved by three top officials: Deng Xiaoping, Hu Yaobang and Wan Li. It agreed that: (a) the Daya Bay project would go ahead; (b) the Guangdong provincial government would work with Hong Kong and the Macau Affairs Office (港澳办公室) on the issue; (c) Beijing would let people in Hong Kong know about Deng's determination to continue the Daya Bay project; (d) Li Peng would approach the British and French to explain the Chinese position on nuclear energy; and (e) the government would inform the public about nuclear energy technology and demystify it.

The decision made by the CCP had two components. First, even though the government was determined to push ahead with the nuclear project, it would be patient with the opposition and the public and was ready and willing to explain and justify the project. There was a general belief that those opposed to the Daya Bay project were poorly informed and they would support it if they were better informed about the safety record and benefits of nuclear power plants. Second, the government had no intention to undermine the confidence of people in Hong Kong about their future, especially that of its educated citizens, whom the government was 'keenest to cultivate for the sake of the territory's future stability and prosperity.'[26] Meanwhile, the government would not tolerate anyone using the opportunity to launch a political campaign against the Sino-British agreement to return Hong Kong to China in 1997 or the political system on the mainland. To accomplish these objectives, the government sent scientists and policy makers to speak at public arenas to different audiences in order to explain the policies.

The official line on environmental protection was announced later that year: 'the chief points in protecting the environment involve site selection, disposal of nuclear waste, and strict control over any release of radioactive materials.'[27] There was no intention to stop the nuclear project, but the government did emphasise the importance of building a safe nuclear power station. Newspapers carried articles with headlines such as *Please don't be afraid, the Qinshan nuclear facility is safe and reliable*. The instruction from the central government regarding the Qinshan project was 'to give primacy to assurances of safety and adhere to the principles of "safety first" and "quality first",' according to an article in *Liao Wang* (瞭望), an official magazine of the CCP.[28] The government restated the responsibilities of the newly created NNSA – to enforce 'the newly-promulgated safety regulations on the siting, design, operation and quality assurance of the Qinshan and Daya Bay nuclear power projects.'[29]

Li Peng, then vice-premier in charge of energy and technology development, toured the Daya Bay site just days before delegates from Hong Kong went to Beijing to deliver the petition of the protestors to shelve the project. His message was that 'safety first and quality first' must be the guiding principle in the construction and operation of nuclear facilities. Meanwhile, to inform the world that China was serious about its nuclear energy programme despite the Chernobyl disaster and the demonstration in Hong Kong, the vice-minister of Nuclear Industry told the media that nuclear power would gradually become the country's second primary energy source.[30]

Scientists, especially those holding government positions, underlined the merits of nuclear energy with the belief that disseminating information about its advantages would undoubtedly be sufficient to convince people to support the project. 'The fission of nuclear fuel in the nuclear power stations can release a large amount of radioactive waste that could harm all living things if it escaped,' explained Jiang Shengjie. The technology used at nuclear power plants in most countries, however, was mature enough to prevent serious accidents involving radioactive leakages.

> No one has died of radiation leakages, and even during the 1979 Three Mile Island accident in the United States, which was caused by human error, no one in or outside the power station died. A final analysis showed that people living within 80 kilometres of the Three Mile Island power station absorbed only 1.6 milirems (mr) of radiation per capita, less than the radiation dosage one is exposed to when wearing a luminous watch or watching colour TV for a year.[31]

Other than some mid-ranking cadres, no high-level officials in the central government expressed their views on the issue. A Hong Kong legislator commented: 'The less often they speak out, the more carefully they seem to be weighing the different points of view... if circumstances change, it is always possible that central-level leaders may say the opposite things.'[32]

From the outset, the issue over the Daya Bay project was entangled with the political issue of the day – the return of Hong Kong to China in 1997. The central government was determined to discuss the two issues separately and was reassured by the elite in Hong Kong to support the project. Sir Lawrence Kadoorie, CEO of the China Light, repeatedly told his counterparts in China that 'he was aware that the public was wary of nuclear energy, but added that this was probably the result of a lack of understanding.'[33] Sir Y.K. Pao, a Hong Kong business tycoon, simply said: 'Nuclear power is worth a go. One cannot stop eating because of the possibility of choking.' Scientists rallied behind the project too. Chenning Yang, a Nobel Prize-winning physicist, argued against an 'attitude of hysterical rejection' of nuclear power, and stressed the need to build the plant to world standards and to 'supervise people who monitor the safety operations.'[34] Some members of the Legislative Council also supported the project on the basis that since there were hundreds of nuclear power plants worldwide and most were operating safely; they could not see why China could not have one. At the same time, they called for better information to be made available to the public. 'Nuclear power is an extremely complex subject', some contended. 'There is no point in objecting to it without knowing what you are rejecting.'

This was also the line Chinese leaders were taking. If the anti-nuclear movement erupted because of an uninformed public worrying about the safety of nuclear plants, the best way to deal with these public concerns was to publicise the message on the advantages of nuclear energy. The Chinese government was not to change its mind on the project, but would try to convince the public in Hong Kong that nuclear power was safe and clean.

The anti-nuclear movement in Hong Kong soon slipped into the background, partly because of the coalition formed between the business community in Hong Kong and the governments in Guangdong and Beijing, and partly because of the rapid economic liberalisation in China from which Hong Kong benefited the most. It was also because the public, traditionally blasé about politics, was not ready to make this project a political issue yet, which was the intention of the influential Teachers' Union, the newly elected members at Legco and other organisations which challenged the political legitimacy of the CCP and the government in Beijing, and demanded its promise of 'one country with two systems' after Hong Kong's takeover in 1997. Ironically, this anti-nuclear episode changed the political atmosphere in Hong Kong. Since then mass mobilisation has been part of the political life. This experience confirmed the experience in Taiwan and South Korea, where nuclear energy issues eventually triggered democratic movements.

Public debates

While nuclear issues might have become a catalyst for social and political movements in Hong Kong, in mainland China, it was long held that the public would accept, and even embrace, nuclear energy development so long as they were fully informed about the details of the technology. Indeed, the public did not react to the issues associated with nuclear development until the 2000s, partly because so many issues demanded immediate attention and partly because public opinion was not reported until the mid-1990s. Since then, both the electronic and print media have actively involved in reporting some of the serious social and environmental problems. 'The greening of the Chinese state' – 'as is visible in the proclamation of an impressive body of environmental laws and regulations and the strengthening of the environmental bureaucracy'[35] – has opened the window of opportunities for civil society to get organised on environmental issues, however they choose to define them. The internet and mobile phones have heralded profound change in the available channels for airing popular discontent and radically transformed domestic politics.

Think tanks have mushroomed and they are much more willing to speak out than before. An increasing number of social groups have been formed in pursuing their causes, including environmental protection. The emerging middle class often brings some of its concerns into open discussion. The influence of these groups on decision making varies. Yet the fact that some of these players have managed to make some NPP projects an issue debated publicly signifies the fundamental change of political life in China. Scientists and many public officials are convinced that public acceptance of nuclear projects is not only a matter of public perception of nuclear technology risks but also of 'complex social, cultural and historical factors,'[36] and therefore they need to be dealt with rather than brushed off.

In general, the public has shown support for nuclear energy development in China.[37] The Institute of Nuclear and New Energy Technology, Tsinghua University, conducted surveys on public opinion on nuclear energy expansion between 2002 and 2006. Their findings show about 80% of those surveyed supported nuclear energy expansion. The rate of support was higher in 2004–06 than it was in 2002–04. Another survey conducted at the same time shows that more than 76% of survey participants had only heard of nuclear energy in general terms and had no further knowledge. This finding was in contrast to the general argument held in many OECD countries that lack of knowledge about nuclear energy is the key to the opposition against nuclear development, because this lack of understanding often subjects the debates to emotion rather than reason. The public might have limited knowledge of nuclear energy, but they do know the consequences of coal-fired electricity generation and are suffering from environmental pollution as the result of heavy reliance on coal.

Surveys conducted by INET asked questions on four separate issues:

- *Opinions on benefits*: (1) benefit to the national power supply; (2) benefit to lower electricity prices; (3) benefit to environmental protection.
- *Judgement on risks*: (4) judgement on operational risks of nuclear power plants; (5) judgment on the risks of nuclear waste (nuclear proliferation is not yet included here).
- *Knowledge of nuclear power*: (6) how much nuclear knowledge the public has; (7) self-assessed familiarity with nuclear power.
- *Trust in the parties concerned*: (8) trust in government agencies; (9) trust in nuclear experts.[38]

The finding of these surveys shows that 'the benefit to power supply, the safety judgment on nuclear power plants, the trust in experts and benefit to environment protection ... are more influential in the public acceptance of nuclear power.'[39] The other side of the coin is that a lack of understanding of other factors might not constitute an important influence in shaping the public's view on nuclear development. To be more specific, the apparent consensus on nuclear energy expansion in China can be explained by the widespread concerns about environmental pollution and energy shortages.

Improving environmental conditions and providing adequate electricity supplies are more pressing challenges than the fear of nuclear disaster.[40]

Furthermore, China currently generates only a fraction of its electricity from NPPs. To the public, any risks or uncertainties are only 'potential'. Also since China started its nuclear energy programme relatively late, nuclear waste is not yet a major concern. The technology used is relatively new and none of the NPPs is near recommission.

The consensus on nuclear energy expansion, however, does not necessarily translate into support for specific NPPs and their locations. When asked whether they would support a nuclear power station in their own region, less than 50% of the people surveyed gave a positive answer, which is in line with the feeling in many other countries that have developed nuclear power plants. This is, however, more than a 'not-in-my-backyard' syndrome, with concerns about potential risks and desolate landscapes; it is about the livelihood of those living around the potential sites of nuclear power plants, the issue of resettlement or new employment opportunities. They can have equally important impacts on policy making. Resettlement, for example, for hydro projects in the past several years has literally forced the government to halt the proposed projects.[41]

'Media liberalisation is widely, though not universally, regarded as a precursor of political liberalisation in authoritarian states.'[42] Media liberalisation includes that in newspapers, magazines, radio, television as well as internet usages. Media liberalisation often takes place in parallel with the emergence and expansion of a middle class and societal organisations. As a scholar argues, 'Civil society and internet energise each other in their co-evolutionary development in China.'[43] In China, the number of radio and television outlets expanded throughout the reform period: it jumped from 38 to 541 in the 1980s, and then it swelled to 3200 by 2006. The number of newspapers and magazines published in China also expanded to 1900 and 9700 respectively.[44] The development and expansion of internet and mobile phone usage have allowed small groups of activists and even individuals 'to exercise influence disproportionate to their limited manpower and financial resources.'[45]

Many social organisations have been formed across China. In October 1993, *China Daily* reported that there were about 1500 autonomous groups operating on a national level and an additional 180 000 operating locally. By the end of 2008, 230 000 were registered with the Ministry of Civil Affairs. Among them, 6716 were identified as environmental organisations.[46] In general, these environmental groups have received strong support from the Chinese media mainly because 'environmental issues are newsworthy, loaded with moral and political meanings and policy implications, yet politically safe because they fall in line with the state policy of sustainable development.'[47]

Environmental issues are controversial in all societies because they are multi-faceted subjects that have political, social and economic components.

Disagreements erupt not because people do not want to protect the environment, but because they cannot agree on which sections of environment should be given what priorities. Nuclear power has been advocated as a clean energy and most people in China have accepted the argument in its favour. Yet, when a site is chosen in an area along the coast, those whose life depends on fishing worry about the impact on fish stocks and those visiting beaches for holidays worry about potential risks or simply do not want to see the scenery spoilt or the destruction of the natural environment. This is not only a perception issue. NPPs can have a direct impact on life. Some nuclear scientists at INET, Tsinghua University, concluded after a laboratory experiment that, with six units of 100MW nuclear power project, the ocean water used for cooling the reactors would increase its temperature by 4°C within 1.13km^2, by 2°C within 6.5 km^2, and by 1° within 22.6 km^2. Changing water temperature will have direct impacts on fish stocks.

As the country develops and the average income rises, Chinese citizens want comfortable transportation, and wish to upgrade their housing and enjoy their vacations. Improved economic conditions often make them more vocal about what they consider as socially and environmentally acceptable. Information technology helps them spread the message and their grievances. Their interests are not always in line with those who still have to struggle to maintain a decent life. The political landscape has changed significantly in China. Public opinion must be taken seriously in decision making. This is the broader context in which the controversies over the proposed NPP in Haiyang (海阳) in Shandong province have arisen.

In the early 1980s, the Shandong provincial government created an office to look into the possibility of a nuclear project in the province. Shandong then had an independent grid that was not connected to those in its neighbouring provinces. There was and is no hydro potential and more than 99% of the generation capacity in the province was coal-fired thermal. NPPs became an attractive option.

In the mid-1990s, CNNC conducted a preliminary feasibility study in Lengjiazhuan (冷家庄), Haiyang and Hongshiding (红石顶) in Rushan (乳山) county, less than 50 km away from each other along the coast. Peasants and fishermen living in the area recalled:

> In the mid-1990s, many people, including some foreigners, came here with all sorts of equipment. We thought they were trying to build a port here. Later when we were told it was to be a nuclear power plant, we did not know what to make of it. Then word came that it was dangerous and poisonous and with a nuclear power plant to be built here field rats could grow as large as human beings. We had no idea what to believe.[48]

Nothing further happened until the early 2000s. In 2002, power shortages spread across the country and Shandong suffered shortages not only of generation capacity but also shortages of coal supplies. With a population of 94 million, Shandong had a total generation capacity of about 44GW (84% of that in Australia with a population of 21 million). Its electricity consumption per capita just reached the national average, but it was only 67% of that in Beijing. Because 99% of its generation capacity was coal-fired thermal, Shandong needed to 'import' more than half of the coal used for power generation from other provinces. 'When you rely on coal imports, you are a hostage of others', stated some local officials.

The provincial government pushed hard to get a couple of nuclear power projects on the national plan. In 2005, it became clear that Shandong would get three projects. In Rongcheng (荣城), China would build its first 20MW high-temperature gas-cooled reactor-pebble bed module (HTR-PM) demonstration project. The second project was the Haiyang (海阳) NPP. In 2003, NDRC approved the construction of two units of 1000MW at Haiyang with space enough to build another four units at the same site. The construction of the project did not officially start until 4 June 2009. One of the difficulties was to convince fishermen in the area to relocate. The local government wanted the project, and it sent its officials to every household: 'each of us needed to convince at least three households to move', recalled a local official. It was not easy. In one typical case, it was reported, an old woman in her 60s, sitting in front of her house, told the official: 'My parents lived here and were buried here; so were their parents and their parents before them. I am not going anywhere'. The official literally sat in her house for three days, waiting for a positive answer.

Some fishermen drove local officials away as they heard what they were there for. After they were offered compensation of 30 000 yuan for each person (including children), many agreed to move, but they often regretted their decision soon after. One fisherman, who had moved 7 km inland said: 'It is only seven kilometres, but without direct access to the sea, what is the use of our boats anymore? Yes, I have five people and received 150 000 yuan, but I have lost my way of life and had to sell the boat.'[49] The construction also affected the lives of those who did not have to move. After their stories were reported in various media outlets, some of them received compensation from the Haiyang project funds.

In the Medium- and Long-Term Nuclear Energy Development Plan (2005–20), the NDRC also listed another site in Shandong, Hongshiding (红石顶) in Rushan (乳山) county, less than 50 km away from the Haiyang project. The three projects are within 120 kilometres. This site would host 6x1000MW units, planned to be built in the 12th FYP (2011–15), as part of the ambitious national plan to build a capacity of 60–70GWe by 2020. Word about the site came out, however, long before the Plan was made official. Newspapers and magazines published articles and questions were raised: (a) could Shandong accommodate a total 12GWe nuclear capacity by 2020? (b) What would be the environmental consequences with the project? Those in charge of electricity development in both central and the provincial government argued that 'there would never be a surplus of generation capacity, given that the country was in such shortage of the capacity at the peak of its industrialisation and urbanisation'. The public accepted the argument without great disagreement. The issue of environmental impact, however, was a completely different matter.

Two issues were at the centre of the debate: the safety of the power plants and the potential degradation and damage to beaches. Regarding safety concerns, for those working in the nuclear industry, the technology used in current NPPs is safe and the public needs to be informed with correct knowledge. CNNC, its subsidiary in Shandong and other shareholders, launched a campaign about the advantages of nuclear energy and the safety of nuclear power stations. They even sponsored organised trips for local peasants and fishermen to visit Qinshan I, II and III in Zhejiang.

On 18 March 2006, the SEPA issued the 'Interim Method of Public Participation in the Environmental Impact Assessment', in which public participation is defined and explained in detail. It required that those responsible for major construction projects with potentially adverse environmental impacts or which affect the public in any way hold seminars, broadcast the views of experts and conduct public consultation before they receive an environmental impact assessment.

Investigation and public communication were conducted in 2006 and 2007. The site was around a small fishing village. There were no big factories nearby and there was a natural deep harbour that had easy access to deep-sea fishing. The area had never been hit by a typhoon or cyclone in its entire 280-year recorded history. The village had about 600 people with 200 fishing boats. In addition to traditional fishing, they have also recently developed fish farms of shrimps and sea-cucumbers, a delicacy in China, Japan and South Korea. The average annual family income was 100 000 yuan, which was quite high compared with other families.

It was reported that feasibility studies of nuclear power sites had been undertaken in the region in 1995 and locals had expected something would happen in the near future.[50] Indeed, when their living standard was low in the 1990s, they had wanted large projects to be constructed in the area. In the 2000s, living standards for fishermen improved a great deal. Some still expected that the government would come in and develop the region and they would move on to another kind of life. One fisherman, who had invested 5 million yuan in a 40-acre fish farm, responding to a reporter's question, said: 'We need power plants; even if the current government would not build a nuclear power plant now, it would have to build one in a couple of years. It is an energy security issue.'

Others were not this understanding. They raised many new issues, including compensation, resettlement, future job opportunities, and so on. A proposal was put forward that a new village would be built along a harbour, but that it would not have the same easy access to the sea. For some fishermen, this meant that they would have to outlay more on diesel oil to go fishing, which would reduce their income. Some villagers wanted better housing and better compensation for their relocation and losses. Some expected the new project would bring the economic boom to the region and they wanted to secure the job opportunities before the project started. Many wanted to know: (a) who would pay for their resettlement and other losses, such as those of sea cucumber and other exotic fish farms; (b) who would pay for the increased fuel expenses of their fishing boats; (c) whether their children would be guaranteed a job at the nuclear power station and (d) whether it was safe to live next to a nuclear power station.

In an area of 5 km of this proposed site, there were about 10 000–15 000 residents (not including tourists) who had quite different concerns and interests. In the early 1990s, the provincial government in Shandong proposed the development of fishing villages into tourist areas around Rushan (乳山). The area had been known in some quarters as 'Hawaii of the east' and in 2002, the National Tourist Administration of the central government listed the place as 4A tourist resort. Developers rushed in and invested heavily in infrastructure, marinas, golf courses and apartment buildings. The apartments were then sold to the urban middle class from Beijing and Tianjin and people from interior provinces such as Shanxi and Inner Mongolia as their retirement or vacation homes.

When it became known that one of the three sites for NPPs in Shandong would be at Hongshiding, next to the developing beach resorts, residents asked: (a) would it be dangerous to live there after the construction of a nuclear power station; (b) would their property depreciate in value; (c) would seawater and seafood still be safe in the future and (d) would the project affect tourism in the region and therefore their potential income? Seasonal residents also wanted their beach view protected.

'Throughout world history, every advance in communication technologies has made it easier for insurgent groups to organise collective action and harder for states to prevent them from doing so.'[51] China is no exception. Media liberalisation, and in particular the growth in internet and mobile phone usage, helped mobilise those who opposed the project. It was through these media that the public became aware of the issue of nuclear energy development and the potential risks to the coast.

The founder of one of the first non-governmental organisations in China on environmental protection, The Ocean Protection Commune (大海环保公社), proposed the use of its website as a platform to collect signatures from those opposing the project in the name of protecting the environment. According to its website, the commune was a loosely organised club with only 30–40 volunteers. Its reputation spread quickly because of the activities it organised, but the website organisers became disappointed when they realised that most of the 5000 signatures collected were not from the locals but from developers and owners of apartment buildings. 'They (the developers) were the ones who destroyed the beaches and environment first and now they are the ones who are opposing the project, not the fishermen,' said a website spokesperson in an interview. 'I am not against NPPs; I just do not think they should be built in this beautiful beach and so close to residential areas.'

Other groups, such as the Nuclear-Free Silver Beach Forum (银滩无核论坛) and the World's First Beach (天下第一滩) were formed or offered their support to the opposition. From summer 2006 onwards, SEPA started receiving complaints and petitions from these groups and few of the complainants were from local people. The main complaint was that developers had already invested heavily in the area and it was too late for them to pull out. NPPs in close proximity to their developments would ruin their investment.

SEPA organised surveys on how the public within a distance of 80 km of the site viewed the construction of a NPP in the region. When the survey was carried out in Weihai, 65 km from the potential site, responses to the survey were surprising: 'The site is so far away, why do you come here and ask about our response?' asked one respondent. 'Since you are here, it must be dangerous; otherwise you wouldn't have come to collect the public response,' said another. This raised the serious issues of how to find the best balance between winning public support without spreading fear. The media faced similar challenges too, as they were reporting the conflicts without presenting nuclear power plants as real safety and security concerns.

In January 2008, *China Business Journal* (中国经营报) ran a four-piece series 'Competing Interests over the Rushan Nuclear Project: Who Represents the Interests of Local People'? A pertinent point raised in the article was: who exactly should be classified as 'local people'? Were they fishermen who had lived in the area for generations, developers who had invested heavily, seasonal residents who had purchased apartments but only spent their summers there, or the small minority who had recently moved to the area because of its development? Seasonal residents accounted for more than one-sixth of the total population and they insisted they should have the same input in the debate as local residents and fishermen, who were a small minority. In mid-December 2007, the local environmental agency of the county government called for a meeting and a vote was cast: four to one in favour of the nuclear project. The developers and associations representing property owners accused the government of abusing their power by holding the meeting at a time when most of them were not present and their interests could not be represented.[52]

In sum, this new form of communication is vastly different to the traditional means of mass mobilisation in that it is faceless. Those who 'signed' the petition and who gave their opinion on the websites did not live in the area and this: (a) undermined their cause, and (b) stirred up resentment from local people who felt that their concerns were not being heard and that their agenda was being 'stolen' by 'city boys'. Consequently, in the eyes of both of the governments in Beijing and Shandong, and the nuclear industry, the opposition had little legitimacy and therefore debate over the Rushan Nuclear Project could be ignored. This does not, however, mean that the government was willing to push the project through regardless. The positions of the central and local governments differed greatly.

The local government was eager to get this project online. A total investment of 60 billion yuan for the project was too great a temptation to resist: 'We need this money for education, our hospitals and other local developments, which developers or seasonable property owners would not be willing to pay,' argued local government officials. A nuclear project would mean more jobs and opportunities not only for the local economy but also for the nuclear power companies.

The central government, nonetheless, had no intention of rushing the project through, partly because at the time inflation was rising and partly because it had its own reservations about having three nuclear projects (two of them would be 6x1000MW) in such close proximity to each other. In 2007, when the Medium- to Long-Term Nuclear Energy Development Plan was officially published, NDRC stated that the site at Rushan would need further study.

The industry was pressing hard to get the project. In October 2005, CNNC and the Shandong government signed an agreement on the preliminary framework for nuclear power plant development and CNNC also sent some people to the area to carry out preparatory work. In May 2006, the provincial government established a Preparatory Office for the Shandong Hongshiding Nuclear Power Limited Corporation. In November 2006, a formal agreement was signed in the provincial capital about the contribution of stakes of the company: CNNC would contribute 51%, Luneng (鲁能, a provincial unity company) 33%, Huadian (a national utility company) 10% and another local company would contribute 6%.

In December 2007, SEPA issued a statement: 'Rushan nuclear project was not submitted to the SEPA for environmental studies and the agency was not asked to assess the site or the project.' This was a preventative measure because a major problem for projects deemed illegal (whether for investment or for environment reasons) was that enterprises often went ahead with projects before they requested approval or started projects before they received approval from the regulatory agencies (as the Chinese would say 先斩后奏 – kill first, report afterwards).[53]

The media kept the story alive and the Internet provided outlets for both sides, especially the opposition, to express their views, even though it is difficult to judge how many people were against the project and whether their views were considered as part of the decision-making process.[54]

'Offensive' or 'defensive' environmental protection

Nuclear energy creates environmental concerns. 'Nuclear energy production requires infrastructure that gives rise to broadly similar issues to other industrial and power generation facilities, such as land-use, thermal emissions and potential chemicals (or other) pollutants.'[55] The mining and milling of uranium produces radioactive waste; nuclear power generation produces spent fuel that contains various degrees of radioactive waste; and decommissioning nuclear power plants involves 'decontamination of structure and components, demolition of components and buildings, remediation of contaminated ground and disposal of the resulting waste.'[56] These productive activities can be termed 'offensive' environmental consequences. Scientists have tried to develop technologies to mitigate the consequences, for example, by building up temporary or permanent storage places for radioactive wastes or reprocessing the spent fuel from power generation. In other words, the offensive environmental consequences of nuclear energy production are 'neutralised' by shifting radiological impacts from one location to another and by shifting the impacts from the present to the future.

'The first line of defence against environmental damage is, of course, the prevention of nuclear accidents by continuing reinforcement of nuclear safety programs.'[57] Governments provide tight regulations for 'defensive' environmental consequences of nuclear production from potential incidents, accidents and disasters to protect the environment and citizens. The public responds to the two types of environmental consequences in quite different ways. People seem to be more concerned about the 'defensive' environmental consequences, the location and safety of nuclear power stations than the 'offensive' consequences that may affect the environment and humankind in the future. Temporary 'neutralisation' of the environmental impacts by shifting the burden from the present to the future may raise ethical and moral debates in other countries, especially developed ones.

In China, the optimistic view about humans' capacity to develop new technology to deal with nuclear waste or handle decommissioning still predominates. Indeed, rarely has environmental pollution associated with nuclear energy, such as uranium mining, radioactive waste or decommissioning, entered the public debate. They are considered topics for scientific communities to discuss rather than the public, partly because the issues are not yet priorities and partly because there is an optimistic assessment, as science and technology develop, that nuclear waste problems will be resolved. After all, most nuclear power plants have a lifespan of 40–60 years and the first nuclear power station came online in China in 1994.

In this context, the public wants assurance from the government that nuclear power stations are safe and secure. Commenting on the safety culture and capability of maintaining a safe nuclear operating system, Professor Andrew Kadak at MIT said:

> Given the numerous stories of deaths and injuries in China's coal mining industry, there is the perception that all industries in China are operated in the same manner. This is not the case. China's commercial nuclear

plants are subject to inspections by the World Association of Nuclear Operators (WANO) and are under the International Atomic Energy Agency's safeguards inspections. Using international indicators, the WANO inspections provide the Chinese operators with an assessment of how their operations compare to other nuclear power plants in the world. In general, the inspections show that the reactors are operated in conformance with international protocols and expectations.[58]

Philippe Jamet, the director of the division of nuclear installation safety at the IAEA, said that China had welcomed foreign inspectors at its reactors and that 'they show pretty good operations safety.'[59] From the time plans for nuclear power plants were first put on the table, there was a consensus between the nuclear industry and the government that the fundamental principle for China's nuclear energy development has to be 'safety first, quality first'. The safety of every nuclear power station is crucial for the nuclear industry, not only in China but worldwide. This was one of the main reasons that, soon after it made its decision to start a nuclear energy programme, the Chinese Government took a dual approach. In 1982, a group of experts was put together to study and draft the necessary safety regulations. With little knowledge of the subject matter and regulations, the Chinese team adopted the safety laws and regulations of the IAEA.

In October 1984, the central government created the NNSA as an independent regulator to supervise and manage safety regulations, and appointed Jiang Shengjie, a nuclear expert, as its first director. NNSA was to 'draft the basic laws regarding the use of atomic energy, to formulate safety regulations, guidelines and standards for civil nuclear facilities, to establish strict, effective procedures for safety approval of Chinese made and imported civil nuclear facilities, to issue manufacturing permits and operating licences, to examine and monitor the safety of civil nuclear facilities already approved or now in operation, to coordinate research efforts of the state departments and local authorities and to undertake international exchange and cooperation concerning nuclear safety.'[60]

The second aspect of the two-track approach was that China joined the IAEA in 1983, placing its nuclear programme under the IAEA safeguard provisions. As Qi Huaiyuan, a director at the Chinese Ministry of Foreign Affairs stated, 'If China joins the IAEA, it will accept the relevant provisions in the statute of the agency, including the relevant provision on safeguards.'[61] In 1985, it volunteered its civil nuclear facilities for international inspections.

For Qinshan I, most work was done by the Chinese. Operation and maintenance procedures and instructions were established and reviewed by Chinese designers and engineers. It was inspected by the IAEA officials several times during its construction and for its pre-operation preparation. On 3–21 April 1989, for example, an IAEA team of international experts from Canada, France, Germany, Italy, Japan, Spain and the US reviewed management of the project, quality programmes, civil work, mechanical works, operation preparation and training of operational personnel. They examined construction work as well as actual operation. To ensure its safety, the Chinese hired the experienced American Sargent & Lundy Engineers, which was established at the turn of the 20th Century, to conduct an overall review of the design of its 17 main and supplementary systems. A key part of the design of reactor pressure containers was then examined a second time by Italian experts. Sargent & Lundy Engineers were also hired to draft the safety analysis report.[62]

For its first commercial nuclear power station, Daya Bay, GNPJVC engaged American Bechtel Company for quality assurance (QA). It also commissioned EDF to undertake overall technical responsibility. To ensure the international standard of operation and management, GNPJVC signed a contract with EDF under which more than 115 Chinese engineers were sent to France to receive various forms of operation and maintenance training, for no less than a year. Overall, GNPJVC transplanted French practices and procedures in operation and maintenance, and on the station management side an experienced French plant manager was appointed to head the Daya Bay power station.

From the start, those involved in the nuclear energy programme understood that NPPs could not be designed and constructed without codes and standards. They followed IAEA codes and basically referred to US codes and standards. They were willing to engage foreign companies in applying these codes and meeting standards. Yet, the Chinese government's decision to expand nuclear generation capacity quickly 'stirred up a lot of concerns' because, some argue, a safety culture has not yet been established in China, and because of the lack of qualified regulatory staff. In late October 2009, Prime Minister Wen Jiabao ordered a quintupling of the safety agency's staff (to 1000) by the end of the next year, according to the US regulators. IAEA accepted a Chinese request to send a team of international experts to the country the following year to assess staffing and training. 'They don't have very much staff, when you compare their staff with how many they will need,' commented Phillippe Jamet from IAEA.[63]

Conclusion

All nuclear projects raise concerns associated with the environment. Scientists and nuclear industries tend to emphasise the contribution nuclear power makes to the environment because it replaces or supplements thermal power generation capacities. The public, on the other hand, tend to be more concerned with potential releases of radioactivity in routine or accidental conditions, radioactive waste management and disposal and proliferation of nuclear weapons. There is general support for nuclear energy development in China, not because the public knows a great deal about nuclear energy but rather because the environmental consequences of its coal-fired generation capacity are widely felt. There is also a consensus among scholars at universities, think tanks and the nuclear industry that before a nuclear power plant is approved and built they, along with the government, must try to convince the public that nuclear energy is safe, clean and good for the general development of the country. People are no longer willing to go along with the government's policies blindly; if they are not fully informed they can easily be manipulated by a few who are motivated for political reasons, especially in this information age. 'Transparency' is the best way to deal with public fear and to win the support of the public who are sceptical of the way the government handles risk.[64]

Without sounding condescending, a senior member of the Chinese Academy of Sciences emphasised that the public fear of nuclear power plants was natural, even though it might not be justified, and the best way to deal with it was to inform people about nuclear energy development. This would not only legitimise nuclear programmes but also place the nuclear industry under public scrutiny and would, in turn, help ensure safety and quality of the nuclear energy programme.[65] These arguments might be taken for granted in democratic societies,[66] but they represent a marked political change in China. Governments, both central and local, have opened various communication channels when nuclear energy projects are a concern to the public and have not shown any intention of suppressing those who oppose them.

Notes

1. Fewsmith and Rosen, 2001, pp.151–174.
2. Fountain, 1979.
3. Olson 1965; Ostrom 1990.
4. Baum, 2008, p.174.
5. Shanghai Revolutionary Committee, 'Rations on Electricity Consumption', 10 April 1978; Beijing City Government, 'Provisional Plan for Electricity Rationing', 17 August 1983. Both documents can be found at the SPCC's homepage: http://chinapw.cep.gov.cn.
6. World Bank and UNDP, 'Energy Sector Management Assistance Program', No.101/89. Washington, DC: The World Bank, May 1989.
7. World Bank, *China: Long-Term Development, Issues and Options*. Washington, DC: The World Bank, 1985, p.5.
8. Li Yingxiang, 'Nuclear Technology for Peaceful Purposes', *China Today*, April 1991, pp.67–68.
9. Zhao Ren-kai, 'Present Status and Prospects of Nuclear Power Development in China', in 'Proceedings of the 9th Pacific Basic Nuclear Conference, Nuclear Energy, Science & Technology Pacific Partnership', Sydney, Australia, 1–6 May 1994, p.60.
10. Richard P. Suttmeier and Peter C. Evans, 'China Goes Nuclear', *China Business Review*, September–October 1996, p.18.

11. Cooper, 1999; Sun, 1996; Zhong, 2000.
12. 核电发展站略研究编委会, 1999, pp.12–13.
13. Kevin Platt, 'China's Nuclear Power Program Loses Steam', *The Christian Science Monitor*, 21 July 2000.
14. IEA, *Key World Energy Statistics*, 2005–2009. Paris: OECD.
15. Zhang 2007, p.3547; IEA, 2006a; Berrah, et al., 2007. World Bank and SEPA, 2007.
16. World Bank, *Clear Water, Blue Skies: China's Environment in the New Century*. Washington, DC: The World Bank, 1997; World Bank, *The Little Green Data Book*, Washington, DC: The World Bank, 2007.
17. Wang and Lu 2002, p.8.
18. NEA, 2008, p.121.
19. 国家发展和改革委员会, 核电中长期发展规划, October 2007.
20. Yee and Wong 1987, p.618.
21. Larry Jagan, 'Political Fall-Out from Hong Kong's Nuclear Power Plant', *Economic & Political Weekly*, 21:40, 4 October 1986, p.1737.
22. Yee and Wong 1987, p. 624; 'A Nuclear Hong Kong'? *Asia Week*, 12:29, 20 July 1986, p.29.
23. 'Daya Bay: A Cloud of Doubt', *Asia Week*, 12:26, 25 June 1986, p.16.
24. 'New Storm Over Daya Bay', *Asia Week*, 12:36, 7 September 1986, p.20.
25. 'Daya Bay: The Gloves Come Off', *Asia week*, 12:39, 31 August 1986, p.24.
26. 'A Nuclear Hong Kong'? *Asia Week* 12:29, 20 July 1986, p.27.
27. Zhou Zhumou, 'Ensuring Nuclear Safety', *China Construction*, 35:8, August 1986, p.23.
28. Meng Fanxia, 'Progress of Construction of Qinshan Detailed', *Liaowang [Outlook]*, 7 July 1986, pp.14–15, in 'China Report: Economic Affairs', JPRS-CEA-86–108, 9 October 1986, pp.68–72.
29. Wu Jingshu, 'France to Transfer Advanced Safety Technology "Free of Charge"', *China Daily*, 21 July 1986, p.1.
30. Zhou Zhumou, 'Ensuring Nuclear Safety', *China Reconstructs*, 35:8, August 1986, pp.20–26.
31. Jiang Shengjie, 'Developing China's Nuclear Power Industry', *Beijing Review*, 18 June 1984, p.19.
32. Quoted from 'A Nuclear Hong Kong'? *Asia Week* 12:29, 20 July 1986, p.31.
33. Chan Chi-keung, 'Nuclear Power "Key to Protecting Environment"', *South China Morning Post*, 19 September 1989, p.4.
34. 'A Nuclear Hong Kong'? *Asia Week*, 12:29, 20 July 1986, p.28.
35. Ho, 2007, p.197.
36. Liu, et al., 2008, p.2834.
37. Sheng Zhou and Xiliang Zhang, 'Nuclear Energy Development in China: A Study of Opportunities and Challenges', *Energy*, article in press (2009).
38. Liu, et al. 2008, p.2834.
39. Ibid, p.2837.
40. China Atomic Information Network, 'Public Opinion Decides the Nuclear Development', September 5, 2006, at http://www.atominfo.com.cn/newsreport/news_detail.aspx?id=044FE6BC6.00000042.0029&contype=NC.
41. Mertha, 2009; Yang and Calhoun, 2007.
42. Baum, 2008, p.161.
43. Yang, 2003, p.405.
44. Baum, 2008.
45. Chase and Mulvenon, 2002, p.xii.

46. Saich, 2000; Ministry of Civil Affairs of PRC, '2008 Statistical Report of Civil Affairs', at http://cws.mca.gov.cn/article/tjbg/200906/20090600031762.shtml, accessed on 28 February 2010.
47. Yang, 2005, p.56.
48. 何伊凡, '村里要建核电站', 中国企业家, 28 July 2008, at http://www.cnemag.com.cn/zzlm/baodao/nengyuan/2008–07-28/173974.shtml; Mcng Bin, 'Shandong to Set Up Nuclear Power Station', *Beijing Review*, 38:43, 23–29 October 1995, p.27.
49. Ibid.
50. Meng Bin, 'Shandong to Set Up Nuclear Power Station', *Beijing Review*, 38:43, 23–29 October 1995, p.27.
51. Shirk, 2007, p.103.
52. '乳山核电站多方利益博弈: 谁能代表当地的民意'? 中国经营报, 14 January 2008.
53. See, for example, IEA, *China's Power Sector reform: where to next?* Paris: OECD, 2006; 崔民选, 中国能源发展报告, 北京: 社会科学文献出版社, 2007.
54. A website sponsored by the China Electricity Association carries several media reports on the debates. See http://np.chinapower.com.cn. Another site is: http://www.sina.com.cn, accessed on 28 February 2010.
55. NEA, 2008, p.140.
56. Ibid, p.261.
57. Sam Emmerechts, 'Environmental Law and Nuclear Law: A Growing Symbiosis', *Nuclear Law Review*, 82:2, 2008, p.99.
58. Andrew C. Kadak, 'Nuclear Power: "Made in China"', http://web.mit.edu/pebble-bed/papers1_files/Made%20in%20China.pdf, accessed on 1 March 2010.
59. Keith Bradsher, 'Nuclear Power Expansion in China Stirs Concerns', *New York Times*, 15 December 2009.
60. 'Nuclear Safety Bureau Established', *Beijing Review*, 27:47, 19 November 1984, p.7.
61. 'China to Join Atomic Energy Agency', *Beijing Review*, 26:34, 22 August 1983, p.9.
62. Han Guojian, 'China: A Country of Nuclear Power', *Beijing Review*, 34:51, 23–29 December 1991, pp.12–15.
63. Keith Bradsher, 'Nuclear Power Expansion un China Stirs Concerns', *New York Times*, 15 December 2009.
64. See, for example, 雷润琴, '我国核电建设的舆情分析与对策', 环境保护, 80, 2008, pp.63–65.
65. Articles published in *People's Daily* in 2004, 2005, 2006.
66. Irwin, 2006, pp.299–320.

9
Is Nuclear the Future?

Is nuclear energy indeed the future? It depends: the IAEA, IEA and the Chinese government would like to emphasise that nuclear power is efficient, reliable, clean, safe and large enough to be used as base-load to, if not solve, at least, alleviate the pressures from the twin challenges China and the world are facing – energy security and climate change. Others see the current move as mere 'nuclear amnesia' because nuclear power will not be able to meet the growing demand or cut carbon emission sufficiently to make a dent in the two main problems, especially in China. One way of describing it perhaps is: 'nuclear power alone won't get us to where we need to be, but we won't get there without it.'[1]

China plans to build 40–60GWe nuclear generation capacity by 2020. Globally, this capacity would be sufficient to satisfy the total electricity demand of a high-consumption country, such as Australia, with carbon-free electricity. For China, 40–60GWe would provide only 4% of the electricity generated then, a minute portion, well below the world's average of 16%. For Guangdong and Zhejiang where limited energy resources are available and most nuclear power stations are located or being built, 15–20% of their electricity would be from nuclear power stations by 2015.

The challenges China is facing are real: while coal currently provides nearly 70% of its energy, it is depleting quickly. At current rates, even if China meets its nuclear target, its total carbon emission will rise by 72–80% by 2020. Something has to be done. China is too large to have a single solution to the problems. Nuclear along the coast, wind and solar for the west and interior, and improved energy efficiency overall might keep China and the world out of imminent energy and climate disaster in the near future.

Can China meet its target of nuclear energy development after so many failures in the past 30 years? Again, it depends. It can be done because other countries have done so before. China now has 8.4–9.1GWe in operation and 20.9GWe under construction. To meet the target, it would have to build 2 or 3 units every year in the next decade. At its peak, the US managed to build 4.63 units a year. France built 13 units in 1976–80 and another 24 units in

1981–85, an equivalent of 3–4 units a year. The construction span was 67 months. If China could match this speed, it would have 30GWe by 2015 and might be able to reach 40GWe by 2020 with another 20–30GWe under construction. The French experience shows, however, after the peak (1976–85), it took much longer to construct nuclear power plants with an average of 86 months in 1986–90, 93 months in 1991–95 and 124 months in 1996–2000. This slower development can be explained by increasing public concerns about the safety of nuclear power after the Chernobyl disaster. It could also be the fact that the electricity market reached its saturation stage.

In contrast, in Japan, the construction span shortened considerably after its initial 61 months in 1976–80, to 47 months in the 1980s and 44 months in the 1990s. China's current record is 73 months for the first three units finished in the 1990s, and followed by 60 months for the next 6 units in 2001–05. The last two units of Russian reactors took 80 months to complete. This average 71-month construction span is longer than an average of 60 months in South Korea but faster than most countries with nuclear power stations. To achieve what France, Japan and South Korea have done in their nuclear development needs other conditions too.

It needs sufficient financial capital available for nuclear development. From 1978 (when economic reforms began) to 2008, China's real GDP grew at an average annual rate of 10%. From 1980 to 2008, China's economy grew 14-fold in real terms; real per capita GDP grew over 11-fold. By some measures, China is now the world's second-largest economy.[2] Both its fiscal and monetary conditions were healthy when it decided to expand its nuclear power capacity. This means that it probably could spare the financial resources and foreign exchange to launch a large nuclear programme. When the global financial crisis hit in 2008, the Chinese government put in place a stimulus package of 4 trillion yuan (an equivalent of US$586 billion). Initially, 45% of this package was designed to go to infrastructure. An adjusted allocation gave infrastructure 38% of the stimulus package, an amount of 1.5 trillion yuan, over US$200 billion. The size of the package shows that if the central government is determined to expand its nuclear energy programme, it could spare the financial resources.

The changes of allocating the stimulus package also show that there are and will be competing demands for financial resources. With per capita income less than 13% of that in the US and 18% of that in Japan, China's 800 million poor people will have quite different needs from the 400 million so-called middle class. The core of the financial question is not whether the country can afford to build 30–40 units of nuclear power plants; rather it is the question of maintaining some degree of equity – equity between the richer and more developed coastal regions where nuclear power plants probably will be built and poor interior provinces whose demand for financial resources is even greater, equity between the nuclear and other energy sectors, and equity between those whose electricity consumption is reaching the level of middle-income countries and those 8 million people in remote rural areas without access to electricity at all.

A rapid development of nuclear energy depends on standardisation and localisation of technology. For now, the Chinese government seems to have decided that PWR is the model the country will adopt. There are four types of PWRs competing for being the base model – AP1000, CNP1000, CPR1000, and EPR. The first three models are all developed from the Westinghouse reactor. Japan has managed to localise two models of technology – BWR and PWR – with 30 units of BWR and 24 units of PWR in operation. The US has adopted two models too with PWR accounting for two-thirds while BWR the rest. The international experience shows that it is not too late for China to decide its technology route. The current research and development on HTR is promising and offers a safe and smaller size option for areas with high population concentration. But, surely a choice has to be made to deliver the economy of scale, the speed of production and safety of the nuclear industry.

The nuclear fuel market is global and with the international community seeking a multilateral approach to nuclear fuel supplies, China has the options of producing by itself and importing from others. Its pursuit of the back-end of nuclear fuel cycle technology – spent fuel reprocessing and fast breeder reactors – is motivated more by a requirement of a distant nuclear future and political prestige as a major global player than immediate need. It is a question how resources will be allocated and what priority will be given to the advanced technology development.

Increasingly, policy entrepreneurs have become part of the policy-making process in China. Their ability to shape the issues subjects nuclear energy development increasingly to the demands of public opinion. The traditional argument in this area states that if the public is well informed, they will accept nuclear as a viable solution to energy security. An argument applied specifically to China is that 'when the public knows nothing about nuclear technology or its dangers, there's not going to be any fear of broad-based opposition to new plants.'[3] As the nuclear industry expands, the public probably will get to know enough to be scared, but not sufficient to be reassured. As the public is no longer a passive taker of policies, and the media and internet help spread their concerns, the government has to deal with the public opinion and popular concerns in making nuclear policies. Meanwhile, their trust of government officials is low, but their trust of civil society groups is even lower.[4] This may create an opportunity for cooperation between the government and the nuclear industry to legitimise nuclear energy development in local contexts. It is unlikely the government will impose a decision on the public regarding nuclear development.

More than anything else, the nuclear future in China depends on how politics evolve. In 1980, Deng Xiaoping proposed, 'Energy is an issue of primary importance in economic area'. In 2010, energy issues were once again given the priority when the State Council created an overarching agency, the National Energy Commission, headed by the Premier. It remains to be seen whether this institution would work, but it is clear that the Chinese government has been aware that energy became a complex issue and could not be managed without a centralised agency to coordinate in formulating strategy and planning development. This was the problem highlighted by a Western reporter in 1981: 'A final cause of China's energy crisis is poor coordination, planning, and management.'[5]

This 'poor coordination, planning, and management' was often explained as the product of the 'fragmented institutions' in the central government, competing interests of ministries, bureaucracies and provinces, 'unbalanced influence' between the weak government agencies and powerful corporations', or the weak central government vs. rich provinces. It is indeed the combination of all these developments that is the cause of the inability of the Chinese government to develop a consistent long-term strategy for nuclear development and thereby contributes to the consistent failure to meet the targets the government-set over the past 30 years.

None of these issues is unique to nuclear energy development. The government's lack of capacity in policy making and implementation is identified in energy, environment and many other policy areas. Building this capacity, however, requires more than merely creating a centralised decision-making authority. Policies are not only about choices. They are institutions themselves. Altering the configuration of organisations is only one aspect of institutional change. Changing the rules of the game and the expectations takes time. Tracing the history of nuclear energy development in China allows us to understand the forces behind it and identify the opportunity and possibilities for progress. Since new initiatives introduced to address contemporary demands add to, rather than replace, pre-existing institutional forms, we cannot expect a quick fix of the capacity to formulate the long-term strategies and stable policies that are so essential to nuclear energy development. In turn, we cannot expect a rapid expansion of nuclear energy programmes in China, not because of the economic and technical difficulties, rather because of the lack of institutional capacity to ensure such an expansion.

Notes

1. Mazen M. Abu-Khader, 'Recent Advances in Nuclear Power', *Progress in Nuclear Energy*, 51, 2009, p.225.
2. Wayne M. Morrison, 'China's Economic Conditions', Congressional Research Service, 7–5700, RL33534, 11 December 2009.
3. Kevin Platt, 'China's Nuclear-Power Program Loses Steam', *The Christian Science Monitor*, 21 July 2000.
4. Teets, 2009.
5. Clark, 1981, p.48.

Selected Bibliography

Amsden, Alice. 1989. *Asia's Next Giant*. New York: Oxford University Press.

Anderson, Jonathon. 2007. 'Solving China's Rebalancing Puzzle', *Finance and Development*, 44(3): 32–35.

Andrews-Speed, Philip. 2004. *Energy Policy and Regulations in the People's Republic of China*. Hague: Kluwer Law International.

Baum, Richard. 2008. 'Political Implications of China's Information Revolution: The Media, the Minders, and Their Message', in Cheng Li (ed.) *China's Changing Political Landscape*. Washington, DC: The Brookings Institution Press, pp.161–84.

Berg, Sanford V and Tschirhart, John. 1988. *Natural Monopoly Regulation*. New York: Cambridge University Press.

Berrah, Noureddine, Feng, Fei, Priddle, Roland and Wang, Leiping. 2007. *Sustainable Energy in China*. Washington, DC: The World Bank.

Blackman, Allen and Wu, Xun. 1999. 'Foreign Direct Investment in China's Power Sector: trends, benefits and barriers', *Energy Policy*, 27(12): 695–711.

Byrne, John and Hoffman, Steven M. 1996. *Governing the Atom*. New Jersey: Transaction Publishers.

Camilleri, Joseph A. 1984. *The State and Nuclear Power*. Seattle: University of Washington Press.

Campbell, John L. 1988. *Collapse of an Industry*. Ithaca: Cornell University Press.

Caro, Robert A. 1982. *The Years of Lyndon Johnson*. New York: Vintage Books.

Chan, Adrian. 2003. *Chinese Marxism*. London: Continuum.

Chan, Joseph Man and Lee, Chin-Chuan. 1991. *Mass Media and Political Transition*. New York: The Guilford Press.

Chase, Michael S. and Mulvenon, James C. 2002. *You've Got Dissent!* California: Rand Corporation.

Cheung, Tai M. 2009. *Fortifying China*. Ithaca: Cornell University Press.

Chow, Daniel C.K. 1997. 'An Analysis of the Political Economy of China's Enterprise Conglomerates', *Law and Policy in International Business*, 28(2): 383–433.

Chubb, John E. 1983. *Interest Groups and the Bureaucracy*. Sanford, CA: Stanford University Press.

Conroy, R.J. 1989. 'The Role of the Higher Education Sector in China's Research and Development System, *China Quarterly*, 117: 38–70.

Cooper, Deborah E. 1999. 'The Kyoto Protocol and China', *Georgetown International Environmental Law Review*, 11(2): 401–37.

Diamond, Douglas W. 1991. 'Monitoring and Reputation', *Journal of Political Economy*, 99(4): 689–721.

Dow, Stephen and Andrews-Speed, Philip. 1998. 'Considerations for Foreign Investors in China's Electricity Sector', *Oil and Gas Law and Taxation Review*, 16(8): 311–17.

Downs, Erica S. 2000. *China's Quest for Energy Security*. Los Angeles: Rand Corporation.

——. 2004. 'The Chinese Energy Security Debate', *China Quarterly*, 177:21–41.

——. 2006. 'The Brookings Foreign Policy Studies Energy Security Series: China', The Brookings Institution, December.

Dumbaugh, Kerry and Martin, Michael F. 2009. 'Understanding China's Political System', Congressional Research Services, R41007, 31 December.

Elliott, David. 2007. ed. *Nuclear or Not?* Hampshire: Palgrave Macmillan.

Esposito, B.J. 1991. 'Energy Policy after the Thirteenth Party Congress', in King-yuh Chang (ed.) *Mainland China after the Thirteenth Party Congress*. Boulder: Westview.

Evans, Peter. 1995. *Embedded Autonomy*. Princeton: Princeton University Press.

Falk, Jim. 1982. *Global Fission*. Melbourne: Oxford University Press.

Feigenbaum, Evan A. 2003. *China's Techno-Warriors*. Stanford: Stanford University Press.

Fewer, H. and Altvater, W. 1977. 'Technology Transfer by Industry for the Construction of Nuclear Power Plant', *Annals of Nuclear Energy*, 4(6–8): 235–48.

Fewsmith, Joseph. 2008. *China since Tiananmen*, New York: Cambridge University Press.

Fewsmith, Joseph and Rosen, Stanley. 2001. 'The Domestic Context of Chinese Foreign Policy: Does "Public Opinion" Matter'? in David M. Lampton (ed.) *The Making of Chinese Foreign and Security Policy in the Era of Reform, 1978–2000*. Stanford: Stanford University Press, pp.151–74.

Fiancette, G. and Penz, P. 2000. 'The Problems of Financing a Nuclear Program in Developing Countries' in IAEA, *Nuclear Power in Developing Countries: Its Potential Role and Strategies for Its Deployment*. Vienna: IAEA, pp.153–61.

Frankenstein, John. 1999. 'China's Defence Industries: A New Course'? in Mulvenon, James C. and Yang, Richard H. (eds) *The People's Liberation Army in the Information Age*. Santa Monica, CA: Rand.

Frankenstein, John and Gill, Bates. 1996. 'Current and Future Challenges Facing Chinese Defence Industries', *China Quarterly*, 146: 394–427.

Fridley, David. 1992. 'China's Energy Outlook' in US Congress Joint Economic Committee (ed.) *China's Economic Dilemmas in the 1990*. New York: M.E. Sharpe, pp. 495–526.

Friedberg, Aaron L. 2006. '"Going Out": China's Pursuit of National Resources and Implications for the PRC's Grand Strategy', *NBR Analysis*, 17(3): 5–34.

Gallagher, Kelly. 2006. 'Limits to Leapfrogging in Energy Technologies? Evidence from the Chinese Mobile Industry', *Energy Policy*, 34(4): 383–94.

Gallagher, Michael G. 1990. 'Nuclear Power and Mainland China's Energy Future', *Issues and Studies*, 26(12): 100–20.

Gill, Bates. 2001. 'Two Steps Forward, One Step Back' in David M. Lampton (ed.) *The Making of Chinese Foreign and Security Policy in the Era of Reform, 1978–2000*. Stanford, CA: Stanford University Press.

Gittings, John. 2005. *The Changing Face of China*, Oxford: Oxford University Press.

Grimston, Malcolm C. and Beck, Peter. 2002. *Double or Quits?* London: Royal Institute of International Affairs; Earthscan.

Haggard, Stephan and McCubbins, Matthew D. 2001. *Presidents, Parliaments, and Policy*, New York: Cambridge University Press.

Harding, Harry. 2006. 'China Goes Global', *National Interest*, 85: 57–66.

Hecht, Gabrielle. 1998. *The Radiance of France*. Cambridge, MA: The MIT Press.

Ho, Peter. 2007. 'Embedded Activism and Political Change in a Semi Authoritarian Context', *China Information*, 21(2): 187–209.

Hong, Eunsuk and Sun, Laixiang. 2006. 'Dynamics of Internationalisation and Outward Investment', *China Quarterly*, 187:610–34.

Hu, Weixing. 1994. 'China's Nuclear Export Controls Policy and Regulations', *The Non-Proliferation Review*, 1(2):3–9.

Huang, Yasheng. 1996. *Inflation and Investment Controls in China*. New York: Cambridge University Press.

Hunt, Sally. 2002. *Making Competition Work in Electricity Markets*. New York: John Wiley.

Irwin, Alan. 2006. 'The Politics of Talk: Coming to Terms with the "New Scientific Governance"', *Social Studies of Science*, 36(2): 299–320.

Johnson, Chalmers. 1982. *MITI and the Japanese Miracle*. Stanford: Stanford University Press.

Johnston, Alastair. 1995/96. 'China's New "Old Thinking"', *International Security*, 20(3): 5–42.

Jones, Dori. 1981. 'Nuclear Power: Back on the Agenda', *China Business Review*, 7 January–February: 32–35.

Joskow, Paul and Parsons, John E. 2009. 'The Economic Future of Nuclear Power', *Daedalus*, 138(4): 45–59.

Kadak, Andrew C. 2005. 'A Future for Nuclear Energy', *International Journal of Critical Infrastructure*, 1(4): 330–45.

Kan, Shirley and Holt, Mark. 2007. *US-China Nuclear Cooperation Agreement*. CRS Report for Congress.

Kursunoglu, Behram N., Mintz, Stephan L and Perlmutter, Arnold. 1998. (eds) *Environment and Nuclear Energy*. New York: Plenum Press.

Lampton, David M. 2001. (ed.) *The Making of Chinese Foreign and Security Policy in the Era of Reform, 1978–2000*. Stanford, CA: Stanford University Press.

——. 2008. *The Three Faces of Chinese Power: Might, Money and Minds*. Berkley: University of California Press.

Lardy, Nicholas. 1992. *Foreign Trade and Economic Performance in China, 1978–1990*, Cambridge: Cambridge University Press.

Lewis, Joanna I. 2007. 'Technology Acquisition and Innovation in the Developing World' *Studies in Comparative International Development*, 42(3–4): 208–232.

Lewis, John W. and Litai, Xue. 1988. *China Builds the Bomb*. Stanford: Stanford University Press.

Li, Nan. 2006. (ed.) *Chinese Civil-Military Relations: the transformation of the People's Liberation Army*. London: Routledge.

Li, Xiaobing. 2007. *A History of the Modern Chinese Army*. Kentucky: The University Press of Kentucky.

Liang, X. and Goel, L. 1997. 'The Achievements and Trend of Tariff Reform in China', *Utilities Policy*, 6(4): 341–48.

Lieberthal, Kenneth. 2004. *Governing China*, 2nd edition. New York: W.W. Norton & Company.

Lieberthal, Kenneth and Oksenberg, Michel. 1988. *Policy Making in China: Leaders, Structures and Processes*. Princeton: Princeton University Press.

Liu, Changxin, Zuoyi Zhang and Steve Kidd, 2008. 'Establishing an Objective System for the Assessment of Public Acceptance of Nuclear Power in China', *Nuclear Engineering and Design*, 238(10):2834–28.

Lu, Yingzhong. 1993. *Fuelling One Billion*. Rockville, MD: Washington Institute Press.

Mertha, Andrew. 2009. '"Fragmented Authoritarianism 2.0": Political Pluralisation in the Chinese Policy Process', *China Quarterly*, 200: 995–1012.

Miller, Leland R. 2006. 'In Search of China's Energy Authority', *Far Eastern Economic Review*, 169(1): 38–42.

MIT. 2003; 2009. *The Future of Nuclear Power: an interdisciplinary MIT study*. Boston: Massachusetts Institute of Technology.

Mounfield, P.R. 1991. *World Nuclear Power*. London: Routledge.

Murray, Fiona E., Reinhardt, Forest and Vietor, Richard. 1998. 'Foreign Firms in the Chinese Power Sector: Economic and Environmental Impacts' in McElroy, Michael B., Nielson, Chris P. and Lydon, Peter (eds) *Energizing China*. Boston: Harvard University Committee on Environment.

Naughton, Barry. 1988. 'The Third Front: Defence Industrialisation in the Chinese Interior', *China Quarterly*, 115: 351–86.

——. 1993. 'Deng Xiaoping: The Economist', *China Quarterly*, 135: 491–514.

Naughton, Barry and Lardy, Nicholas. 1996. 'China's Emergence and Prospects as a Trading Nation', *Brookings Papers of Economic Activity*, 2: 273–444.

Newberry, David. 1999. *Privitisation, Restructuring and Regulation of Network Utilities*. Cambridge. MA: The MIT Press.

Noble, David F. 1977. *America by Design: Science, Technology and the Rise of Corporate Capitalism*. New York: Knopf.

Nolan, Peter. 2001. *China and the Global Economy: National Champions, Industrial Policy and the Big Business Revolution*. New York: Palgrave.

North, Douglass C. 1990. *Institutions, Institutional Change and Economic Performance*. New York: Cambridge University Press.

Olson, Mancur. 1965. *The Logic of Collective Action: Public Goods and the Theory of Groups*. Cambridge, MA: Harvard University Press.

Ostrom, Elinor. 1990. *Governing the Commons: The Evolution of Institutions for Collective Action*. New York: Cambridge University Press.

Park, Chung-Taek. 1992. 'The Experience of Nuclear Power Development in the Republic of Korea' *Energy Policy*, 20(8): 721–34.

Paterson, Walt. 2007. *Keeping the Lights on: Towards Sustainable Electricity*. London: Chatham House/Earthscan.

Perking, Dwight. 1994. 'Completing China's Move to the Market', *Journal of Economic Perspectives*, 8(2): 23–46.

Pierson, Paul. 1993. 'When Effect Becomes a Cause: Policy Feedback and Political Changes,' *World Politics*, 45(4): 595–628.

Pollack, Jonathan D. 1972. 'Chinese Attitudes towards Nuclear Weapons, 1964–9', *The China Quarterly*, 50: 244–71.

Pollack, Jonathon D. and Reiss, Mitchell B. 2004. 'South Korea: the Tyranny of Geography and the Vexations of History' in Kurt M. Campbell, Robert J. Einhorn, and Mitchell B. Reiss, (eds) *The Nuclear Tipping Point: Why States Consider Their Nuclear Choices*. Washington DC: Brookings Institution Press, pp. 254–92.

Price, Terence. 1990. *Political Electricity: What Future for Nuclear Energy?* Oxford: Oxford University Press.

Rajan, Raghuram G. 1992. 'Insiders and Outsiders: The Choice between Informed and Arm's Length Debt', *Journal of Finance*, 47(4): 1367–400.

Ravallion, Martin and Chen, Shaohua. 2004. 'China (Uneven) Progress against Poverty', World Bank Policy Research Working Paper 3408, September.

Reardon-Anderson, James. 1987. 'Nuclear Power Policy Debate in Mainland China, 1978–86', *Issues & Studies,* 23(10).

Robinson, Thomas W. 1982. 'Chinese Military Modernisation in the 1980s', *China Quarterly*, 90: 231–52.

Saich, Tony., 2000. 'Negotiating the State', *China Quarterly*, 161:124–41.

Serger, Sylvia Schwaag and Breidne, Magnus. 2007. 'China's Fifteen-Year Plan for Science and Technology', *Asia Policy*, 4: 135–64.

Shambaugh, David L. 1987. 'China's National Security Research Bureaucracy', *China Quarterly*, 110: 276–304

——. 1996. 'China's Military in Transition', *China Quarterly*, 146: 265–98.

Shin, Chul O., Yoo, Seung H. and Kwak, Seung J. 2007. 'Applying the Analytic Hierarchy Process to Evaluation of the National Nuclear R&D Projects: The Case of Korea' *Progress in Nuclear Energy*, 49(5): 375–84.

Shirk, Susan L. 2007. *China: Fragile Superpower*. New York: Oxford University Press.

Simon, Dennis F. 1987. 'Modernising Science and Technology in China', *Current History*, 86(521): 249–52.

Smil, Vaclav. 1981. 'Energy Development in China', *Energy Policy*, 9(2): 113–26.

——. 1988. *Energy in China's Modernisation: Advantages and Limitations*. New York: M.E. Sharpe.

——. 1988. *Energy in China's Modernization*. New York: M.E. Sharpe.

——. 2004. *China's Past, China's Future*. New York: Routledge Curzon.

So, Alvin Y. 1999. *Hong Kong's Embattled Democracy*. Baltimore: The Johns Hopkins University Press.

Steinfeld, Edward S. 1998. *Forging Reform in China: the fate of state-owned industry*, New York: Cambridge University Press.

——. 2008. 'Energy Policy' in Yusuf and Saich (eds) *China Urbanises*. Washington, DC: The World Bank, pp.125–56.

Strauss, Julia C. and Saavedra, Martha. 2009. 'Introduction: China, Africa and Internationalisation', *China Quarterly*, 199: 551–62.

Sun, Homer. 1996. 'Controlling the Environmental Consequences of Power Development in the People's Republic of China', *Michigan Journal of International Law*, 17(4): 1015–49.

Teets, Jessica C. 2009. 'Post-Earthquake Relief and Reconstruction Efforts: The Emergence of Civil Society in China?' *China Quarterly*, 198: 330–47.

Vogel, Ezra. 1989. *One Step Ahead of China: Guangdong under Reform*. Cambridge, MA: Harvard University Press.

Wang Dazhong and Lu Yingyun. 2002. 'Roles and Prospect of Nuclear Power in China's Energy Supply Strategy', *Nuclear Engineering and Design*, 218(1–3): 3–12.

Wang, Qiang. 2009. 'China Needing a Cautious Approach to Nuclear Power Strategy', *Energy Policy*, 37(7): 2487–91.

Weil, Martin. 1982. 'The First Nuclear Power Projects', *China Business Review*, 9 Sept–Oct: 40–44.

Winskel, Mark. 2002. 'Autonomy's End', *Social Studies of Science*, 32(3): 439–67.

Woo, Wing Thye. 2007. 'The Challenge of Governance Structure, Trade Disputes and Natural Environment to China's Growth', *Comparative Economic Studies*, 40(4): 572–602.

Wu, Jiaping, Garnett, Stephen T. and Barnes, Tony. 2008. 'Beyond and Energy Deal: Impacts of the Sino-Australian Uranium Agreement', *Energy Policy*, 36(1): 413–22.

Xu Yi-chong. 2002. *Powering China*. Dartmouth: Ashgate.

Xu, Yuanhui, Hu, Shouying, Li, Fu and Yu, Suyuan. 2005. 'High Temperature Reactor Development in China', *Progress in Nuclear Energy*, 47(1–4): 260–70.

Yang, Dali. 2004. *Remaking the Chinese Leviathan: Market Transition and the Politics of Governance*. Stanford: Stanford University Press.

Yang, Guobin. 2003. 'The Co-Evolution of the Internet and Civil Society in China', *Asian Survey*, 43(3): 405–22.

Yang, Guobin. 2005. 'Environmental NGOs and Institutional Dynamics in China', *China Quarterly*, 181: 46–66.

Yang, Guobin and Calhoun, Craig. 2007. 'Media, Civil Society, and the Rise of a Green Public Sphere in China', *China Information*, 21(2): 211–36.

Yang, Ming and Yu, Xin. 1996. 'China's Power Management', *Energy Policy*, 24(8): 735–57.

Yee, Herbert S. and Yiu-Chung, Wong. 1987. 'Hong Kong: The Politics of the Daya Bay Nuclear Plant Debate', *International Affairs*, 63(4): 617–39.

Young, Warren. 1998. *Atomic Energy Costing*. Boston: Kluwer Academic Publishers.

Yusuf, Shahid and Saich, Tony. 2008. (eds) *China Urbanises: Consequences, Strategies, and Policies*. Washington, DC: The World Bank.

Zhang Shiping. 1997. *Party vs. State in Post-1949 China*. New York: Cambridge University Press.

Zhang, Zhong-xian. 2007. 'China Is Moving Away the Pattern of "Develop First and then Treat Pollution"', *Energy Policy*, 35(7):3547–49.

Zhao, Jianping. 2000. *The Private Sector and Power Generation in China*. World Bank Discussion Paper No. 406.

Zheng Yongnian. 2007. *De Facto Federalism in China*. London: World Scientific.

Zhong, Ling. 2000. 'Nuclear Energy: China's Approach towards Addressing Global Warming', *Georgetown International Environmental Law Review*, 12(2): 493–522.

IAEA and OECD references

IAEA. 2006. *Nuclear Power and Sustainable Development*, Vienna: IAEA.

——. 2008. *Financing of New Nuclear Power Plants*, Vienna: IAEA.

——. 2009. *Nuclear Technology Review 2009*, Vienna: IAEA.

IEA. 2006. *China's Power Sector Reforms: Where to Next?* Paris: OECD.

——. 2006. *World Energy Outlook 2006*, Paris: OECD.

——. 2007. *World Energy Outlook 2007: China and India Insights*, Paris: OECD.

——. 2009. *World Energy Outlook 2009*, Paris: OECD.

NEA and IAEA. *Uranium 2007: resources, production and demand*. Paris: OECD.

NEA. 2000. *Reduction of Capital Costs if Nuclear Power Plants*, Paris: OECD.

——. 2004. *Government and Nuclear Energy*, Paris: OECD.

——. 2006. *Forty Years of Uranium Resources, Production and Demand in Perspective*, Paris: OECD.

——. 2008. *Nuclear Energy Outlook 2008*. Paris: OECD.

OECD. 2005. *Governance in China*, Paris: OECD.

——. 2008. *OECD Reviews of Innovation Policy: China*, Paris: OECD.

World Bank. 1993. *The East Asian Miracle: Economic Growth and Public Policy*. Washington DC: The World Bank.

——. 2007. *Cost of Pollution in China: Economic Estimates of Physical Damage*. Washington, DC: The World Bank.

——. 2008. 'Mid-term Evaluation of China's 11th Five-Year Plan', Report No. 46355-CN, 18 December.

Selected Chinese Publication

当代中国的核工业，北京: 中国社会科学出版社，1987. (*Nuclear Energy Industry in Contemporary China*, Beijing: China Social Sciences Press, 1987).

核电发展站略研究编委会, 1999. 核电发展站略研究, 北京: 中国言实出版社, (Editorial Board of Nuclear Energy Strategic Development, 1999, 'Nuclear Energy Development Strategy', Beijing: China Yanshi Press.

孟戈非.2002.未被揭开的谜底: 中国和反应堆事业的曲折道路，北京: 社会科学文献出版社, (Meng Gefei, *Undiscovered Mystery: development of China's nuclear energy program*. Beijing: CASS Press, 2002).

李鹏. 2004. 起步到发展: 李鹏核电日记, 北京: 新华出版社, (Li Peng, 2004. *Towards Development: Li Peng's Diary on Nuclear Power*, Beijing: Xinhua Press).

定军, 2005. '核电棋局', 中国经济周刊, (October), (Ding Jun, 2005. 'Chess Game of Nuclear Energy', *China Economics Weekly*, (October), pp.12–19.

刘国光, 2006. 中国十个五年计划研究报告, 北京: 人民出版社, (Liu Guoguang, 2006. *Research Report on Five-Year Plans in China*. Beijing: People's Press).

于得义, 2007. 华能集团核电战略的问题分析纪建议, 对外经济贸易大学. (Yu Deyi, 2007. 'An Analysis of Huaneng Nuclear Strategy and Recommendations', Master Thesis of The University of International Business and Economics).

国家发展和改革委员会, 2007. 核电中长期发展规划, (NDRC, 2007, 'The Mid- and Long-Term Development Plan of Nuclear Energy').

明茜, 2008, 竞争还是整合? 中核, 中核建设合并尚无定论, 21世纪经济报道, (Ming Qian, 2008. 'Competition or Merger: CNNC and CGNPC', *21st Century Economic Report* (28 August).

冯志卿, 2008. '中广核"以核养核"' 中国投资, 3, pp.66–70, (Feng Zhiqing, 2008. 'CGNPC's Rolling Development', *China Investment*, 3: pp.66–70.

崔民选, 2008. 中国能源发展报告, 2008, 北京: 社会科学文献出版社, (Cui Minxuan, 2008. *Blue Book of Energy: Annual Report of China's Energy Development*, Beijing: Social Sciences Academy Press).

Index